Praise for Love and Other Fish

"In describing *Love and Other Fish* I could use words like wonder, heartbreak, astonishment, clarity, rumination, and understanding. But what I really want to say is just *read this book*. With its beautiful narratives and contemplations, *Love and Other Fish* explores the strangeness of human experience and the strangeness of looking back. How wonderful and immersive to be guided by such a wise and generous writer."—Beth Nguyen, author of *Owner of a Lonely Heart: A Memoir*

"Hannah Hindley is one of those rare writers whose generosity, knowledge, and lyricism take you on adventures of the heart and wilds while also making you feel right at home. *Love and Other Fish* swims among watery lives, including ours, with assurance, vulnerability, and wisdom. These essays are gorgeous."—Christopher Cokinos, author of *Still as Bright: An Illuminating History of the Moon from Antiquity to Tomorrow*

"These astonishing essays braid grief, science, and love so tenderly that the world itself feels newly textured—each fish, each memory, each bright moment asking us to pay better attention to what keeps us alive. *Love and Other Fish* moves with a deep tenderness—an attention to the overlooked and the nearly forgotten—and in doing so, builds a world where memory, love, and survival can swim and splash beside each other."—Aimee Nezhukumatathil, author of *World of Wonders: In Praise of Fireflies, Whale Sharks, and Other Astonishments*

"Science and the mysteries, the measurables, and the ineffables, the learned, and the felt are pulled together and arranged in this gorgeous collection, a musculature built of personal bravery and a sense of oneness with the natural world. A lyrical sensibility married to the precisions of scientific exploration—a profound combination of ways of being, thinking/feeling, and forging kinship with the wider world. I'm thrilled this book will be in the world, especially at a time when we need to be stunned back into love and care and other worlds."—Lia Purpura, author of *All the Fierce Tethers: Essays*

"*Love and Other Fish* moves fluidly between taxonomy and personal vignette, stitching together a beautifully crafted piece collection that asks us who we are, who we've been, and how we are, ultimately, always plural animal selves. It's a work that teaches and reaches for relation, asking readers to consider the pitfalls and possibilities of our gendered bodies."—Julietta Singh, author of *The Breaks: An Essay*

Love and Other Fish

RIVER TEETH LITERARY NONFICTION PRIZE
Daniel Lehman and Joe Mackall, *Series Editors*

The River Teeth Literary Nonfiction Prize is awarded to the best work of literary nonfiction submitted to the annual contest sponsored by *River Teeth: A Journal of Nonfiction Narrative.*

Also available in the River Teeth Literary Nonfiction Prize series:

Off Izaak Walton Road: The Grace That Comes Through Loss by Laura Julier
Aligning the Glacier's Ghost: Essays on Solitude and Landscape by Sarah Capdeville
Disequilibria: Meditations on Missingness by Robert Lunday
What Cannot Be Undone: True Stories of a Life in Medicine by Walter M. Robinson
The Rock Cycle: Essays by Kevin Honold
Try to Get Lost: Essays on Travel and Place by Joan Frank
I Am a Stranger Here Myself by Debra Gwartney
MINE: Essays by Sarah Viren
Rough Crossing: An Alaskan Fisherwoman's Memoir by Rosemary McGuire
The Girls in My Town: Essays by Angela Morales

Love and Other Fish

Essays on Wonder, Loss, and Transformation

Hannah Hindley

University of New Mexico Press | Albuquerque

Printed in the United States of America

Library of Congress Cataloging-in-Publication Data
Names: Hindley, Hannah author
Title: Love and other fish / Hannah Hindley.
Other titles: Love and other fish (Compilation)
Description: Albuquerque: University of New Mexico Press, 2026.
Identifiers: LCCN 2025054295 | ISBN 9780826369703 paperback |
ISBN 9780826369710 epub
Subjects: LCSH: Hindley, Hannah | Naturalists—Biography | Women authors—
Biography | Creative nonfiction | BISAC: LITERARY COLLECTIONS / Essays | LITERARY
COLLECTIONS / Subjects & Themes / Animals & Nature | LCGFT: Autobiographies |
Creative nonfiction | Essays
Classification: LCC QH31.H642 A5 2026 | DDC 508.092--dc23/eng/20260112
LC record available at https://lccn.loc.gov/2025054295

Founded in 1889, the University of New Mexico sits on the traditional homelands of the Pueblo of Sandia. The original peoples of New Mexico—Pueblo, Navajo, and Apache—since time immemorial have deep connections to the land and have made significant contributions to the broader community statewide. We honor the land itself and those who remain stewards of this land throughout the generations and also acknowledge our committed relationship to Indigenous peoples. We gratefully recognize our history.

Cover illustration by Hannah Hindley
Designed by Felicia Cedillos
Composed in Sabon LT Std

TO THE GIRLS WHO GO LOOKING FOR THEMSELVES,
AND TO THOSE WHO ARE BRAVE IN LOVING WHAT THEY FIND

And when white moths were on the wing,
And moth-like stars were flickering out,
I dropped the berry in a stream
And caught a little silver trout.

—W. B. Yeats, "The Song of Wandering Aengus"

Contents

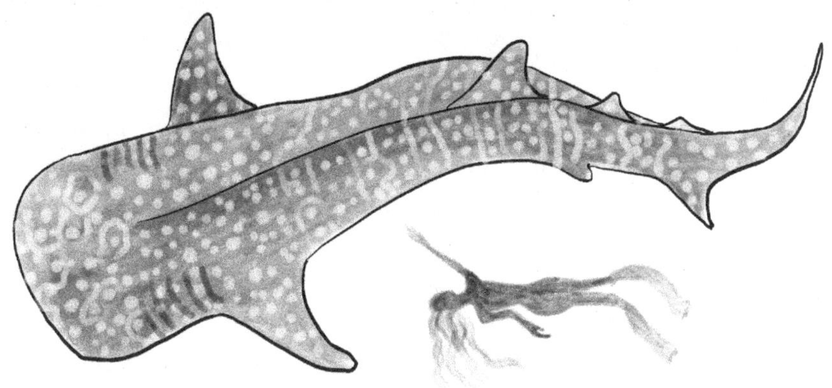

Introduction

WHEN I FIRST MET a whale shark in the Bay of La Paz, I wondered how the boat driver found it—something so huge but findable only by its fin cutting low across the choppy surface. Even after I'd jumped in the sea, where buzzing plankton turned the water gray green, I kicked and spun there in the murk. I looked and looked. It wasn't until the shark—all forty feet of it—came quite close that I could see it. By then, it was so near I could feel the forward push of water as it approached me. Its mouth was wide, and I backpaddled and scooched out of its way, coughing salt water through my snorkel: a rabbit darting from an approaching truck. Though I knew it would not swallow me—knew it did not wish to, even if it could—the enormity of it, and the proximity, set my heart racing. It filled my whole view. It moved like a train. Invisible effort, plunging momentum.

Later, in the deep water off Hawai'i's Big Island, I finally saw a whale shark clearly, free of murk. Out there, two miles from shore, the volcanic seafloor drops six thousand feet into the abyss. Deep currents swirl against the drop-off. The rising water collects strange debris and carries it along in an odd parade. Along that rip, a small boat might come across a crate from Japan, bristling with gummy-necked barnacles, pods of pilot whales with skin like black glass, or a lone whale shark prowling for food in the current. That's where we found her, moving between fishing boats.

I was the only one to jump. The other passengers stayed aboard our small Zodiac, leaning over the edge. The water was shocking—blue in every direction, a prismatic bottomlessness. My heart ran like a river in my ears. But ahead of me and rising up from below swam the shark, and my heart slowed, and I kicked toward her. Her twenty-ton body was as long as a whale's body, gray and dappled with white and slashed with the same blue light that bathed us both.

Later, I'd learn that the patterns on a whale shark's body so resemble the night sky that to better recognize individuals, biologists apply a NASA algorithm designed for identifying star patterns. We overlay what we know about the sky across the bodies of sharks, like layers of tracing paper. Could the same be true, in reverse? We see galaxies on the backs of these giants. But when I look at the dark sky, I see clouds curved like fish bellies and starry freckles across the skin of the night. Can sharks and other fish help us understand the workings of the stars, or of the earth, or even of our own little bodies kicking through the blue? I ask this question, quietly and often, in the pages ahead.

That day off the coast of Hawai'i, a fellow guide arrived with his guests in another Zodiac to look for the shark. He took an underwater picture of me as I swam with her. I still treasure that snapshot. In the photo, the shark is too big to fit inside the frame. She muscles ahead above me, her head out of sight, her body a kaleidoscope of speckles. I lag below, both arms tucked at my side, my whole body taut as an arrow, reaching.

And aren't we all good at reaching, good at looking for connections? We draw lines between stars, build shapes with them in order to tell stories. Like us, fish are patternmakers. Off the coast of Japan, geometric mandalas materialize in the soft seafloor—appearing and melting away, mysterious as crop circles. Only recently have we learned that these are made by male puffer fish, who spend nine days brushing sand into concentric rippled circles, carving out valleys with their fins, decorating the ridges with bits of shell, carting the finest sediment into the center of the circle, soft for nesting, pale as a love letter just before pen touches page. Where I work in Alaska, salmon make long, looping migrations that carry them out to sea, sometimes for years, then back again to cleanly remembered birthplaces deep in bear-musky valleys, or far inland under the glint of summer snow, or in quiet lakes that turn red with the backs of fish that have arrived to close the circle once again. A group of whale sharks is called a constellation, and maybe there's a story there, too.

Anyone who has gone walking with me knows that this collection could just as easily be about beetles, or flowers, or squirrels. I didn't choose to tell these stories because I love fish most of all. Fish are not more exceptional than any of the other little living things that make life in this weird century a little more miraculous, a little more connective.

Still, all my life, fish have risen to swim through my stories: through the burial of my first (and only) pet catfish under the pear tree in our clay-heavy California backyard, and through the shining exhibits at the Steinhart Aquarium, where my father worked when I was small. Through my time as a not-very-good biology student in college, exploring Harvard's sprawling fish collection and learning to take oceanographic measurements at sea from the deck of a sailing ship with the Williams-Mystic Maritime Studies program. Through my windblown winter days guiding in the Sea of Cortez (I still dream of that strange and awe-inspiring morning when a heavy oarfish collided into my body in the shallows off Isla San Francisco). Through the years I've spent as a snorkel guide, following little fish over coral reefs and following big fish into the blue light of the open ocean. Through my time in the desert where, while working toward my MFA, I discovered little topminnows and spikedace populating Arizona's endangered waterways. Through the trackless Alaskan forests that I've spent ten summers navigating, stepping through streams thick with spawning fish and squeezing between colossal trunks of old trees whose wood is made from a thousand generations of salmon.

Even through the life I lead now—unimaginable when I first began writing the essays in this collection—fish swim. In the tide pools on the Sonoma coast, where I leaned close, untethered from time, to watch dappled sculpins in the oppressive early days of the COVID-19 pandemic. In the harissa-rose halibut that my now-husband simmered for me on an early date, to eat in front of the fire as he read to me from *A Natural History of Western Trees*. Fish have moved through so many of the stories that have made me *me*.

Alongside fish, I've lost and loved, flailed and grown graceful, adventured and learned how to hold ground. Again and again and again—like migratory patterns, like constellation lines reaching between stars—fish, it seems, have chosen me.

I once studied under a professor who charted the bodies of fish in colored ink across whiteboards. He insisted that—to our detriment—much of what we think we know about ecology is based on terrestrial life, the things that live on land. For example: Ecologists, in general, believe that the way animals catch and eat their prey requires a special design to match the nature of the prey. I taught this concept, too, when I worked with middle schoolers in the Hoh Rain Forest. We played a game called "bird beak bonanza," experimenting with Darwinian ideas about diversity and specialization. I'd give my students different tools to stand in for beaks—spoons, chopsticks, butter knives, forks—and I'd lay different "foods" in the middle of our circle (beads, marshmallows, marbles). The students were responsible for using whichever tool they'd been given to scoop whatever items they could grab into their bellies, which were small cups.

Not all the utensils could pick up all the foods, but eventually each student discovered which specific item they could grasp with the particular tool they had. They learned with their hands how life evolves toward specialization—different beaks for different foods, just like Darwin's finches, all evolved from a common ancestor. Consider the slender tongue of the anteater, specialized for eating its thousand scuttling breakfasts out of narrow holes in the ground. Or the molars of a deer—flat for munching foliage—compared with a bobcat's jagged carnassial teeth that shear meat like scissors.

Most of what's taught in ecology classrooms is based on these terrestrial examples, and then applied offhandedly to all life, in all places above and below water. Animals diversify in order to take advantage of very different foods, evolving toward specialization, right? Wrong, insisted my professor. What might be fact on land doesn't hold true in the realm of fish.

Where land animals have diversified their feeding mechanisms—molars, tongues, incisors, and so forth—fish have not. Instead of chomping or skewering, most fish, regardless of species, suck. They open their mouths swifter than blinking, and space blooms in the backs of their mouths, and a vacuum yawns open. Watch a goldfish at the surface of its tank. If you're quick enough, you'll see it: how the flakes of food vanish into the fish's mouth with no biting at all. My professor understood that this type of feeding mechanism makes fish versatile. Most of them (unlike the anteater) aren't wedded to a single food source. When the world changes around them, they seek out new snacks. In all of recorded human history, not a single marine species of fish has gone extinct (though not true for their vulnerable freshwater cousins). Most fish resist specialization; they are adaptable, and so they persevere.

All this time, in ecology classrooms around the world, we've been imposing the laws that apply to land life on fish and their ilk. Sure, feeding mechanisms may seem like a niche area of study, but, my professor argued, everything we thought we knew about competition, resource partitioning, adaptation, and niche specialization gets flipped on its head when we actually start paying attention to what goes on underwater. Fish complicate land-centered ecology. They overturn the entire paradigm.

If fish can change the way we consider evolution and ecological relationships, maybe they can be teachers in other ways, too. I look to them as such, in this collection. Though not all these essays are about fish, overtly, I hope that fish surface in places where autobiography falls short and where a different, bigger-than-human perspective might serve to tell a fuller story. Maybe fish are wiggly living clues now, in the Anthropocene—a geologic epoch that we've named after ourselves—that anthropocentrism, or even terrestrocentrism, can't possibly paint a full picture of our ecological place in the world. Move away from the center, the fish remind us. Unclench your fist from around the things you know to be true. If you're looking for answers up there—where wildfires strip land black, where bigger and bigger hurricanes pulverize cities, where pandemic comes walking in from clear-cut edges of jungle, to scratch at each of our windows—move closer to the water. Loosen your grip on yourself, or whoever you believed yourself to be. Step forward. Feel the gut-lift as gravity slackens, the salt water frothing against your hips, your closed lids. Open your eyes. See, there: the murk, the swooshing current cut through with light. See the shark, who has roamed this little planet two thousand times longer than our species has. See how it turns toward you, its flanks like galaxies. See how it opens its mouth, which is as dark and wide as the universe.

Book of Bones

DAD MADE ME MY first pair of stilts in the workshop next to our home, haloed in sawdust and summer light and the spice of calendula wafting in from Mom's garden. He wore no shirt while he worked, and shorts that would embarrass me when I was older—thrifted from the Salvation Army and bright aqua blue and a little too short to be cool. For now, though, I was unconcerned with Dad's fashion faux pas. He was home for the time being, and we were coconspirators, dreaming together of long legs.

The workshop was more cluttered than Narnia, but it was still a journey into some other dimension, to race through the side yard, skip up the broad steps to the open door, and enter Dad's workspace in those stolen weeks when he was home from his port supervisory job two states away. We had a routine: I'd ask for a treasure, and Dad would pause what he was doing and rummage like an archeologist through the shelves and the repurposed tomato boxes in search of just the right thing. We'd reached a serious agreement: Whatever he chose, he'd present it as a true treasure, and I received it as such, with a gasp of delight and feigned awe at the beauty of it. Much of the time, it would be a stupid old nail or a roll of masking tape, though Dad trained me well and I greeted these with gratitude and dazzlement. But sometimes Dad retrieved truly precious things from the workshop's dusty layers. An impossibly delicate miniature rocking chair, crafted from a finely cut and curlicued aluminum can. A machete. A railroad spike. A damiana bottle in the shape of a voluptuous woman. These were fossils of the lives he'd led as an anthropology student in Papua New Guinea, as an electrician for San Francisco's cable cars, as a marine engineer on the *African Neptune*, traveling up the Congo. I knew in my bones that my dad's favorite job was to father me—I spent my childhood in a warm orb of confidence that I was loved more profoundly than any other daughter—but, like the Burgess Shale, my dad contained histories. They glimmered darkly in the strata of his workspace—treasures, each of them.

Now from this catalog of wonders, Dad pulled screws and wooden shafts, Velcro and plywood and a screwdriver. He clamped the shafts together like the long bones of a leg, and he cut the plywood down to size so that my feet could rest atop those long lumber limbs. The bandsaw spat dust; Dad's blue shorts swished synthetically as he moved; the air softened with the scent of sweat and sap. He drilled holes into an old pair of running shoes and attached them to the plywood platforms. He padded the wood around my knees with duct tape and foam. I wasn't supposed to be wearing stilts—the orthopedist had only tentatively given the go-ahead for soccer, let alone circus arts—but maybe my dad understood that the only way to keep me safe was to let me grow strong.

Now that I've begun walking on stilts again, half a lifetime later, I wish I knew what wood my dad crafted that first pair of stilts from. Poplar from the Yard Birds lot in Santa Rosa? Douglas-fir, strong and not too knotty, from the Northern California woods? The stilts are long gone—Mom deemed them too risky after I wore them in our hometown parade: I striding intrepidly, she following in my wake, fretting that someone might dash out and trip me. I cannot ask my dad for his memory of the wood, or for the reason he built those long legs for me, despite the doctor's warning. Like the stilts, he is gone.

Gone things abide sometimes, though rarely. Imagine a fish, huge and toothy, from some long-lost sea. Imagine its armored face, its bus-length body, the turtle-like teeth that self-sharpen as it chomps. It has muscles that can snap its mouth closed as fast as a single video frame on TV, as quick as a nerve signal from fingertip to brain. It slices its way through the watery Devonian.

Imagine the thing that lays it to rest: slow fungal infection, a sudden near-shore mudslide, a volcanic blast, let's say. Its body settles like a retired airplane parked in a field. The ashes from the blast blanket its body. Nothing troubles the remains: the armored scales too thick, the ashes too pillowy and deep. Small things do their small work on the fish's flesh. Bacteria dissolve the soft bits. River muck and ash parade over the place where the fish fell. Weight builds, layer after layer. Mineral seeps through the phyllo folds of sediment, absorbs into the glassy bones, replaces cell and calcium with pyrite and silica until the fallen monster is more rock than fish, enduring long after the last of its living kin have gone.

This is what it takes for a thing to outlast life. Sudden burial. Strong bone.

As for me, my bones haven't always been hard. I mean, this is true for everyone's bones, really. My bones, and yours, and the bones of koalas and whales began as cartilage in the womb. A model of the whole skeleton got laid out in draft form before the bone-building business began. Rubbery rods telescoped open like flowers in the shape of femur, fibula, clavicle. Bone cells encased the soft models, straitjacketing the cartilage. The empathetic body sent blood to feed the dying cartilage, but bone cells hitchhiked in along these new routes and slurped the cartilage away. Tiny canals spider-webbed along the dying cartilage framework. They carried tissue fluid, they connected spongy new bone cells. Calcium salts grew like cave crystals across this scaffolding. The soft bits hardened.

Cartilage held tight where bones connected, allowing flexible places for joints to hinge. Cartilage stayed alive, too, in narrow bands near the bone ends that would allow for limbs to grow as the body grew. Once hardened, bone stops elongating. Without a little cartilage, our skeletons would stay small, brittle, ungrowable.

My bones, though? They were soft in the wrong places. As a toddler, without the speech to name it, I fretted and wept one morning as my parents dressed me for a cousin's wedding. I had spent the morning playing in the neighbor boy's backyard, sliding into his wading pool, prancing in the soft grass: a gentle playdate. Now, I could not be placated. We skipped the wedding, went to the doctor when my leg began to swell. The x-rays showed, improbably, a fractured leg. Such a commonplace morning—no sudden accident, no long fall—and still my little leg had cracked easy as a candy cane.

Again, in preschool, my leg snapped. The boys who were playing with me leapt off the jungle gym, and I followed, and they landed smooth as cats, and I crumpled. *This shouldn't be happening*, the doctor said. Children's bones don't snap that easily, he told us, and certainly not in response to ordinary pressures. He put us in touch with a specialist, and we waited for an opening.

I don't remember the pain of the break, though I hear that broken legs hurt a whole lot. I remember the itch under the cast, the whine of the saw when they cut the plaster free, the way the air filled with chalky cast dust, and the way I shrank away from the saw, worried that they'd cut too deep. I checked my leg for nicks afterward, but the only injury was under the skin, mended imperfectly along the seam of my bone.

I remember watching the fog sift through the fingers of the Golden Gate Bridge as we drove into San Francisco to visit an orthopedist. He fit braces for me, bigger ones each year to keep up with my growing leg. I loved him—Dr. Louie. I brought him art, looked forward to our visits. I would see other children in the clean corridors of the hospital—some in whirring wheelchairs, others with shoes built out on platforms to balance uneven growth—and I wondered if my leg would betray me again. It was weak in the middle, Dr. Louie told us. A congenital condition that made it easier to snap. *Pseudoarthrosis* meant "false joint." Like a knuckle, like a knee, like a wrist, my right tibia was bowed, bendy, breakable. If it broke again before I reached adulthood, it might not grow with me, Dr. Louie explained. My leg would stay small while the rest of my body grew beyond it. *There's no saying if it will get worse, or better.*

Soft things don't preserve well in the fossil record. Earth's history is written in the hard parts of things that were: armor, mandible, limb. To abide, a thing needs stoniness, structure. The past is a book of bones. But life? Life began with softness, with sway. Before seaweed, before flowers, before fingers, the early ocean jiggled with soupy bacteria and single-celled life. Air above the poles sizzled above 120 degrees Fahrenheit, and the seawater was a hot bath. Tectonic plates slammed about, releasing minerals into the sea that little life-forms began to slurp up, to fortify themselves against other small and hungry things. Five hundred million years ago, life started getting creative with these new materials—creatures emerged and disappeared like fireworks: five-eyed *Opabinia*, wormy *Hallucigenia* with spines like stilts, spaceship-shaped *Marrella*. Early almost-fish emerged out of the goo, too, supported by notochords: flexible spines that were not yet bone. These spines strengthened into cartilage, and smooth jawless fish and sharks and other boneless monsters took shape in the long-ago warm sea.

In the human womb, we retrace each sweep of this evolution, like ancestral instant replay. We grow tails and gills, then they dissolve as we melt toward humanness. This is why my undergraduate ichthyology classroom was crowded with pre-med students: doctors look to fish to understand human blueprints, bodies, bones. Just as my cartilage rib cage ossified into bone, so, too, did early fish remineralize themselves into something new. Cartilage draft, bone revision.

Four hundred twenty million years ago, fish began diverging into distinct groups: those who kept their cartilaginous bodies—prehistoric sharks and their ilk—and those who did not. These new bony fish parted ranks, too, diversifying like fingers unclenching from a fist. Ray-finned fish, with diaphanous fins supported by hard spikes, went on to dominate sea and lake. Ninety-nine percent of today's fish are their ancestors: the trout in the whitecapped river, the sea horse nestled in a veil of coral, the goldfish in the bowl. Lobe-finned fish, with fleshy fins secured to their bodies by a single bone, moved toward a very different evolutionary future. Their fins mutated, their joints bent, their bones thickened. They made it possible for life to drag itself, for the first time, ashore. Lobe-finned fish, the fossil record tells us, were the forefathers of legged animals.

I wore a lot of braces growing up, fastened with Velcro, yellowed with use. Each brace came in two pieces—one sculpted for the back of my calf and the scoop of my heel, the other a long lid to be tightened across my socked shin like a shoe tongue. Every morning I strapped myself in and every evening I took the brace off and let my legs swim naked under the bedsheets. Sometimes my leg ached, but only—I imagine—with the ache that all growing legs feel, bone ossifying out of cartilage as they grow longer bit by bit. Mostly, I was restless. *No contact sports*, Dr. Louie advised. *No gymnastics. No skiing.* My bone could not afford to break again. I tried to be good, to love the things that were allowed. I swam. I jumped rope. But when I went away to camp one summer with explicit restrictions, my parents got a postcard from me in the mail:

> *Camp is great. I'm making lots of friends and the food is scrumptious! My favorite activity is stilt walking, and I earned my Certified Stilt Walker card yesterday! So much fun!*
>
> *Lots of love,*
> *Hannah*

What could my parents do? I longed for movement. What I wasn't allowed, I stole for myself.

Fossils reveal the evolution of legs, the patchy path toward land. The bones of lobe-finned fish began to articulate, hinging so that limbs could bend at the elbow. Tiny *Kenichthys*, its skull two centimeters across, swam with pectoral fins that grew in pairs like sets of arms and legs. Four-foot-long *Panderichthys* finwalked its way through shallow waters clogged with fallen branches and rotting plants. *Tiktaalik*, dug up from the Canadian Arctic and dubbed a "fishapod" by paleontologist and writer Neil Shubin, had wrists to resist strong currents, rays like fingers within its fins, and arm bones scarred by long muscles that pulled across them. Its fins had sturdy interior bones, strong enough to for the fish to stilt itself up in the shallows and walk underwater like a four-legged animal. *Tiktaalik*, the fossils say, was the fishy cousin to some long-ago common ancestor of ours that left the water, that evolved down the long millennia into braided strands of family: salamanders, parakeets, elephants, little girls with legs that begged for action.

I got lucky. My legs grew evenly and did not break again. By the time I was in third grade, I was allowed to walk without a brace. The stilts, though, were too risky—*Cut them shorter, at least*, my mom pleaded with my dad, but when he did, they lost their magic. I stubbornly refused to walk on the shortened pegs.

Instead, I danced. I rollerbladed long loops past the horse pastures and vine-blanketed outbuildings of the Sonoma Developmental Center in our little hometown. As a teenager, I went into the High Sierra and learned how to hike for days on end, carrying what I needed in a pack. During college, I found work on an old schooner, teaching kids about the ocean and about themselves. I drew diagrams of fish in chalk on the deck of the boat; I helped a girl with one arm climb the rigging. I pushed northward, where I guided kayaking trips and led hikes through dense Alaskan rain forest where we balanced along fallen trees and waded through quick streams shimmering with baby salmon. I took people into high places in the mountains, where we skated and slid down long talus slopes in search of a good basecamp, loosened fossils clicking underfoot.

The talus slopes scared me. The toppled trees did, too, though I guided people across them with calm know-how. My whole life, I have feared falling—a well-ingrained lesson, I suppose: bone memory, or the constant council of doctors and devoted parents. Still, I stubbornly pushed my way into wild places. Maybe I was making up for lost time. Maybe the momentum of it all was what thrilled: the vertigo of potential, two legs in motion.

In Alaska, years after I took off my brace for the final time, a rock flew from a glacier and crushed my ankle.

I had been exploring an ice cave with a friend—a miraculous cave, different from any other I'd been inside: not blue, like frozen water so often is, but dark, crystalline, glittering blackly in the sweep of our head lamp beams. We were inside the body of the glacier, but we could have been inside another planet. As we climbed back toward the exit, we paused and listened to rocks break loose from the edge of the glacier above us and clatter like poured gravel outside the cave. Ribbons and stalactites of ice caught the light coming in from the outside. We were close to the bigger world, trapped only by rockfall. We crouched in the dark sparkle of the cave and my heart raced and I knew I was unsafe.

Outside, rock cracked against rock.

I waited.

Silence settled.

My friend made a run for it first, and I followed and never saw the boulder coming, only felt the slam of rock against bone, and my foot came out from under me and I couldn't walk and the sun was bright outside the cave and my friend came running. His arm was around my waist, his shoulder was under my arm, and he was pulling me away from the cave, lifting me over the rocks, dragging me toward safety.

If I had been alone, could I have gotten away? Maybe not. The way out was steep and crumbly, and the rocks were falling quickly. I might have lain there, unable to put weight on my ballooning foot, my blackening ankle. Rocks popping loose from the glacier's edge would have pinned my body, broken it in a hundred places, accreted like a heavy blanket. Layers would have collected, slowly. Gray silt from the melting glacier. Ash from taiga wildfires. The soft parts of me would have melted clean of my frame. In the places where water seeped down through the mineral layers, the draft of my bones would have been rewritten, replaced cell by cell by silica, by stone.

But it wasn't. I kept moving, and did not fossilize. So.

What compels us to move despite risk? Answers aren't always written in bone but in softer places. This, for example: a species of skate (small cartilaginous relatives of sharks) possesses the genetic blueprint that allows for the right-left pattern of walking that four-legged animals use. Skates evolved long before bone even existed, let alone limbs. Their "walking genes" coiled into existence long before the first vertebrates ever waddled from sea to shore, the way that movement begins as a thought before fingers go about enacting it.

Before legs ever evolved, the inner instructions to use them did.

Maybe it's in our genes, too—that wiggly urge for movement despite danger, the encoded ache to stretch one limb in front of the other toward some unimagined shore.

It was in the year after my ankle injury that I took up stilt walking again. I was thirty-one years old and emerging from a depressive gloom after months of confinement with my slow-healing limb. A friend invited me to join him for a stilt-walking lesson in a local Tucson park. In the spring light, in the soft grass, my chest thumping with nerves, I strapped into the borrowed legs and wobbled aloft.

The movement came back to me, awkward but smoother with every stride. I lifted my knees, kept my shoulders back.

As the lesson drew to a close, music started up on the other side of the park. We kept our stilts on and followed the thumping bass. Across the way, a pack of local college kids had gathered to celebrate Holi, the Hindu festival of colors. They swirled outward to enclose us. One student cast bright handfuls of chalky dust up toward us, and our faces and shirts and stilts smeared with all the color. *Can we dance with you?* they asked, and we moved closer, stepping to the beat, testing the limits of our long new limbs. I spun in circles, kicking high with one peg foot and then other. I sidled backward, shimmied forward.

And I learned this about stilts: The more you try to hold still, the likelier you are to fall. Rigidity kills. To move on long limbs necessitates constant movement. Walking, it turns out, is easier than standing. Dancing is easier than walking.

The complicated thing about fossils is that bones alone don't tell the full story. High in the Holy Cross Mountains of Poland, where a coral lagoon once ringed the southern coast of the supercontinent Laurasia, geologists dug up muddy tracks encased in stone. They belonged to a large animal, longer than a king-size bed, with footpads nearly a foot across. Whatever it was had evolved beyond "fish." Unlike *Tiktaalik*, it had the strength to support its full body weight on its limbs. It did not drag itself through the shallows. It strode. And the impressions it left in the mud were eighteen million years older than *Tiktaalik*.

Tiktaalik, the celebrated missing link, *Tiktaalik* the amphibious fish, the first tetrapod.

The trackways untidied the story. No longer did we have a clean transition from lobe-fins to lungs to legs, written in stone.

The trackways show that when *Tiktaalik* was still belly crawling its way through mucky shallows, eighteen million years of evolution had already peopled Laurasia with four-legged things that breathed ferny air, that prowled among clicking insects, that moved along the edges of the land on cushioned triangular toes, that two-stepped on long limbs across a muddy Devonian dance floor.

There are no bones to tell that part of the story. The evolutionary instant when fish first grew legs is as soft and mysterious as any long-ago moment recalled through the haze of time. For all its stoniness, the fossil record is just that: Earth's memory, written in soft sediment.

Fossilized bones like *Tiktaalik*'s chronicle only the things that got stopped in their tracks—snapped, buried, preserved. Paw prints reveal something softer, more alive. They show us, faintly, the record of the things that kept moving.

More than twenty years after I walked, braceless, out of Dr. Louie's office for the last time, I spoke with him by phone. I imagined him in his new office, dressed in white, preparing for the morning's surgeries. Maybe his shock of dark hair had started to silver, as my dad's had. Time had passed. Together, we talked about softness. My tibia had a place in the middle of it that wasn't as strong as the surrounding bone, he remembered. But that didn't mean it was too soft—the bone was sclerotic—meaning dense, hard. Bones, Dr. Louie explained to me, are not long, solid broomsticks. They're canalized, woven through with airy latticework. The denser a bone, the more breakable. Stiffness, not softness, was what had kept my bone from being strong.

There are a thousand bone-stiff things in this world that I do not wish to last forever. Border walls, dams, the rigid gleam of a heroin needle at my brother's elbow. Maybe, instead, softness will be how tomorrow's stories take form.

Watch it all sift downward and collect like fine ash—that loose stratum of things too soft for the fossil hunters:

My bendable bones.

The slow-motion dance of animals acquainting themselves with their limbs—across Devonian mud, atop Anthropocene stilts.

A friend's arm, pulling me away from big danger.

The memory of my father, circled in sawdusty light, crafting wooden legs for me, trusting that my own would continue to grow.

Remembering That Life

ISHMAEL LIVED WITH ME for a week before he died in my room. It was January, and cold outside, and he seemed uneasy, restless. Frost had begun to form at the edges of the windows; shut off from the wind outside, the room smelled of dust and salt. Blanketed in my top bunk, I read one last poem out loud that final night, and Ishmael looked at me numbly. His mouth opened, closed. *Sweet dreams*, I whispered, and twisted the switch until the light dimmed and clicked out.

In the morning, strands of hair stuck against my lips from sleeping too close to my pillow, I turned my head to see if he was awake yet. My roommate was at his boyfriend's place, so it was just me in the room this morning, and, as I looked more closely at Ishmael's still form, the dead body of a goldfish.

I guess we've all seen dead fish.

When I was a kid scavenging for driftwood along California's coast, I stumbled across one that had washed up from the deep, eyes glazed, its thin teeth glistening like needles, and I realized I wasn't the only one swimming in those cold swells. I stared in horror for a few spellbound moments before running from it on fast little legs.

At an outdoor market in Seattle one spring, I watched men dressed in glossy foul-weather gear toss silver salmon back and forth, catching the slippery bodies with nonchalant bravado. The fish were almost whole, save for long cuts down their stomachs, out of which the inner parts of them had been torn. Five dollars and ninety-nine cents a pound.

When I got tired of reading Puritan poetry in college, I enrolled in a class that brought us, without an ounce of literary analysis, to an ichthyology collection at the far end of campus. Through a back door in Harvard's Museum of Comparative Zoology, down a white hallway and just past a room crowded with disarranged papers, we found a fish library.

In the stacks: neither the *Encyclopedia of Fishes* nor Kurlansky's *Cod* but actual fish, bottled and lined like books down the shelves.

One and a half million pairs of eyes failed to follow me as I traveled the aisles. Behind their labels, the fish floated in yellowy alcohol, wide-eyed, stiff. The fins on particularly old specimens had begun to shred. Some fish, too large for their jars, pressed against the sides at awkward angles, folded in half or made to curl along the contours of the glass. Others, smaller than my fingernails, drifted in eerie schools together in the same jar, like little pickled hors d'oeuvres. A litter of unborn sharks hovered quietly in a single vessel, their round yolk sacs still connected by a thin cord to their bellies.

Some fish were brown and wrinkled, some spined like pineapples. In bloated places along some of the older specimens' bodies, organs had split through the skin and spilled stilly outward, suspended and grotesque. Other fish had been cleared and stained. We could see every tiny vertebra and fin ray, brittle and bright. Their ribs curved thinly; their jaws were like etched glass. Our whispers sounded loud against the smooth walls.

Intimacy with those little deaths could never have prepared me for the startle of life when I returned from winter break that year to find three real, live fish swimming in a plastic bowl on the table in my living room. My roommate's theater company had played a prank on him by hand delivering four dozen live goldfish to the mail slot in our door. Unluckily, it wasn't he but another, vegan, roommate, who had come back to the room to find wild-finned goldfish slapping and twisting on the wet hallway floor. The thin bag they'd been in had burst in our mail slot just before she arrived. She'd screamed and flung open the door, racing to find containers for them, only realizing after she had stepped on several that the fish had scattered on the inside of the room as well. They'd popped and slid beneath her feet. With panicked hands, she'd scooped up the twenty-five survivors and put them in water. She'd bought them food, but over the next few days, one by one, the battered refugees had drifted into death at the tops of mugs and repurposed bowls.

When I showed up, I did my best to keep the remaining three alive. I knew how gills worked and understood the careful structure of tails—in theory alone—but now I found myself at a real loss. How to feed those little bodies? Where to keep them safe? I bought them a gallon jar at Crate & Barrel, which had thick glass sides rippled like the undersurface of water. I bought food for them (made, I noticed on the package, with the dust of other fish), but when I sprinkled the flakes into their new home, the restless survivors ignored them. Their mouths opened, closed, opened soundlessly. I tensed and knelt down at their level, trying to make sense of their behavior with what little I knew about their anatomy. We stared at one another. A lover of Sondheim musicals, I named them Lee, Harvey, and Oswald. I read to them from Neruda's "Enigmas" before going to bed that night.

> I walked around as you do, investigating
> the endless star,
> and in my net, during the night, I woke up naked,
> the only thing caught, a fish trapped inside the wind.

Mid-verse, a returning roommate opened the door on me. I froze sheepishly. We faced each other with open mouths for a moment before we both broke into laughter—at my absurdity, at his witnessing. *This isn't what it looks like*, I declared. But he smiled, unconvinced, and left me with the open book of poems in my lap and my three disinterested wards busily searching the underbelly of the surface for something I couldn't provide. Feeling exposed, I listened to my friend retreat into his room. I imagine he knew with greater clarity than I why I sat among the fish, why I read to them, why I fought so urgently for their lives.

Before bed, I scattered more flakes into their water, where they floated in thin flecks before saturating and settling, uneaten, among the rocks. Much later, I rose and padded out in the darkness to check on them. Alone in the common room, they drifted slowly in the moonlight, fins spinning restlessly. By morning, Harvey and Oswald were dead.

Or maybe it was Lee and Harvey.

Not knowing, I renamed the final survivor, and Ishmael moved in with me so I could keep a closer eye on him.

In the fish collection, everything was dead, but maybe not

entirely lifeless. Like photos in a darkroom, little histories clung to spine and scale. Following the shaft of an anglerfish's lure—from its base at the top of the head to its knobby, worm-like tip—I could almost see the flashes of bioluminescence scattering as it prowled the abyss, see that bulbous tip twitch as hungry little things wandered too close to the light and lost themselves in the fish's cage of teeth.

The bottled fetuses looked tender, still round with the possibility of birth.

The stained specimens revealed everything. Each transparent torso, with its rainbowed skeleton within, exposed the workings of the life it once held. The sludge of the everyday had been flushed from behind their eye sockets and out from between their invisible muscles, and now they looked like memorials of themselves: not quite art, not quite bodies. They were truths, laid bare. Skeletal, neon purple, they carried their past gingerly in their jaws.

I watched as a teaching fellow laid a small, fresh shark out in a pan and carefully sawed off a slice of its head. The area around its nostrils was cold and a little rough to the touch. The cartilage cut smoothly. Beneath the skin and skull, we found its brain—wet, pink, ridged. It was just flesh now, soft folds of tissue, but I imagined that somewhere among those ridges, its stories were still hidden: the sharp impulses that drove it to snap and eat, the memory of water as it flushed through gills, the final fury of being drawn up with hook and net into the killing air. Written in its skin and in its brain, I could glimpse the movements that carried it through existence. The shark's body revealed its life.

As the knife pressed deeper, the tray filled with dark blood.

Ishmael continued to gasp at the top of his tank, as if he were right on the verge of forsaking his gills and surfacing to taste air instead. I wondered if this was trauma. In his mind's eye, did his two dozen companions thrash lidlessly in the hallway? Could he still feel the sharp passage of air through his gills in those long, urgent moments out of water? Goldfish, I've been told, have three-second memories. I thought about that as I read aloud to him. By the end of each line, had he already forgotten the beginning? By the time he was back in water, could he still remember the close-packed explosion in the mail slot? His anxious mouth seemed too evocative to deny that somewhere in the small chambers of himself, he still held the slap of floor and gasp of companions lodged in some deeper place than mind. I liked to think that even if my words were unintelligible, somehow the overall body of each poem could still wrap itself around him, pressing against bone if not brain, ricocheting from fin to fin in deep, hidden movements.

Our memories are so much more than skull and intellect. Where the mind fails, the physical enters. A rubber band, when stretched between fingers, will remember its original dimensions and sink back into itself when released. The ocean, with vast, circulating momentum, remembers slight changes in wind and weather and carries those motions through the years in broadening circles and deepening currents years after a storm has dissipated. A cleared fish, bright with stain, will maintain its fine-boned shape long after it migrates from wild creek to glycerin bath.

What, though, when the body is gone? In the absence of the physical, does memory stand a chance? Ishmael finally stopped gasping a few days after he moved in. He looked realer than the bottled fish, realer, even, than those sleekly tossed salmon on the other coast. He was neither specimen nor food but something palpable and named and lost. I sang a song in the bathroom, feeling foolish, and flushed him down the toilet.

Two summers ago, my dad and I followed the narrow coastal highway north from our home in California. To our right, rounded yellow hills dipped and rose, punctuated by stray cows and thistle. On our other side, cliffs dropped breath-catching distances down to the sea. New to driving, I sat stiffly behind the wheel, knuckles whitening at each sharp curve.

In the late afternoon, we parked and explored the coast by foot. Ancient rock had lifted up from the sea and rose in terraces in either direction. Wind and tide had licked the cliffs into repeating honeycomb patterns. In the basins and folds where time had carved the shape of itself, water had come and gone, and in its place we found crystallized reserves of salt, spread in gemlike crusts underfoot. Pelican shadows brushed against our faces.

We pitched a tent beneath a pine and watched the orange fan of sunset unfold. In the dusk, we sipped the special red wine saved only for father-daughter excursions and, when the color had dimmed around us and the world grown grainy and gray, we unfolded our sleeping pads and read aloud Ray Bradbury shorts, like in the old days: "The Sound of Summer Running," "The Golden Apples of the Sun." We lay under the sky, darker where the gnarled tree leaned across it, and I asked to hear his own stories. I'd come to learn that he would never share the full arc of his life with me, but instead would only give me isolated memories, the way he'd sometimes bring me a surprise from the cluttered workshop: an old glass bottle, a horseshoe, a bent spring. He'd tell me about the time he learned to build wooden boats in a small town on the West Coast or how, before the days of containerization, he worked as a marine engineer alongside grizzled men who still remembered the days of sail.

Tonight, he told me about his sea voyage to Asia aboard a cargo ship. At a port in Japan, he and a friend found bikes and sped off among the sharp corners and bright noise of the city until twilight found them once more at the water's edge. They got off their bikes and stood in the still evening, watching fishermen cast glowing lures out from the docks. The men called out to one another in a soft language, and my dad listened for the changes in tone and tempo, but couldn't understand their words. He and his friend stood in silence in the gathering night. Set in quiet flight, the little lights skimmed on invisible lines out across the dark water.

At the end of his story, his breath slowed into sleep, and I lay wordlessly, listening to the distant surf crash against arched bodies of rock.

A few weeks later, my dad disappeared during a diving expedition. A friend from work drove me to my hometown that evening and held me with her free hand as I choked back salty fear. I lay curled in my mother's arms that night, and we breathed thin breaths as we waited for the ring of the phone, the knock of the door. When I slept, I dreamed of teams of dark divers circling the place where he'd gone missing. I could almost hear the sound of helicopters in the midnight sky cutting low among islands. The next morning, a policeman mounted our front steps and touched the doorbell once. My mom answered. His body had been found, seventy-five feet below the surface in a kelp bed off Southern California's coast.

He was taken somewhere and cut open, but the autopsy revealed only a healthy, if somewhat waterlogged, body. The Coast Guard found no fault with the diving equipment. For months afterward, the men who had dived with him, guilty, perhaps, that they had surfaced without checking to see if he was following them, evaded speaking to us about the final minutes before my father's disappearance. A mystery, then. Mermaids, maybe, tempting a lost sailor toward death with sweet voices like in the days of old. Or a loss of oxygen.

I chose not to look at him when he was brought north for cremation. Waiting for family to come back out of the funeral home, I realized I would never see his face again: the splayed wrinkles that sunbeamed the corners of his eyes, the wide, whiskered smile that never quite left his mouth even when temper furrowed his brow. I was tempted to change my mind, to run inside and see him one more time as a body before even that disappeared, but a strong sense that he was already gone set an open space yawning in my chest and kept me put. I was in no mood to see that face transformed, emptied. I sat in the car, turned off the music. The light moved slowly across the parking lot.

If memory is a physical retention of things—the energy in a rubber band drawing it back into its original shape, the ocean carrying ancient storms in its vast stomach, an interrupted fetus suspended in an eternal glycerin womb, a bruised goldfish gasping for release as its muscles relive, again and again, the terror of the mail slot—what happens when the rubber band snaps, when the fish is flushed? I washed my hands of shark blood in the Museum of Comparative Zoology and didn't look back to see the animal's chunks of brain slip into the garbage among gill filaments and bile. My dad came home in a bag full of fine gray dust and heavier lumps of material: melted buttons, bone. He is scattered now among mountains, soggy in the leaf litter of forests, drifting like ocean snow in the slow tides off our home coast. Whatever memories his body held have been yielded up to the wind and slow mechanisms of geology.

In the nineteenth century, a young biologist took sponges and tore them apart into individual cells. He set them aside and watched as, with slow, deliberate movements, the cells moved toward one another and came together. They bound themselves into larger formations, first fusing into vague balls, eventually adopting structure. A new body formed from disaggregated scratch. Unified life from torn pieces.

Elsewhere, researchers trained tiny flatworms to respond preemptively to a stimulus by curling up in anticipation of an electric shock rather than stretching out in the presence of light, as is instinctive for them. When the worms were sliced into hundreds of pieces and fed to other untrained worms, the new worms resisted instinct as if they, too, had been trained to fear the shock. They shrank into themselves when the light turned on.

Monarch butterflies cover immense distances as they migrate south to overwinter. Returning north, they only ever reach the lower part of their range before laying eggs and dying. On tremulous new wings, the next generation will complete the northward journey across the continent, closing a careful circle they had never begun. The following fall, in a confident orange flurry, they repeat their parents' journey, gathering for the first time, months later, in the exact trees the last generation found in that circuit too vast to be completed in a single life-span.

Memory, it seems, is not a purely mental exercise, nor is it just a physical record of the bruises and miracles we collect as individuals along the dusty road, but it is a forward movement, a thing that carries histories into new lives and seeps, transformed, under rocks and into wings.

High in the Trinity Alps, I took my mother to a snowmelt creek and waded with her in the quick eddies above a waterfall that my dad and I had hiked to a year before. I carried a sack filled with what was once my father and, as the channel deepened, we cast handfuls of him across the clear surface. The gray flecks soaked through and sank wetly among the smooth pebbles underfoot. Swift fish, mottled with the blues and sheen of the stones around them, brushed against my ankles and slipped again out of sight. The rocks were loud with the smash and froth of falling water, and twilight settled in softly without my hearing it.

We slept there after sundown, head to head, hauled up on a thin island of granite. Mouths opened and closed in the darkness around us: things drinking from the near shore—maybe, inadvertently, lapping up thin flakes of ash with thirsty tongues. I found my mother's hand in the night, warm to the touch, the color of mountains in the dim light. The moon caught in the branches of trees and in the quiet hours that carried us places beneath our own thoughts, distant stars spun slowly in the wide current of the night.

River Time

ON THE SHORES OF Lake Crescent, where mist rises like memory out of the evergreens, Mount Storm King presides. A mountain god—stony-flanked, bearded with fern and cedar—he has always kept order. So the legend goes.

Long ago, a famine came to the people who lived in two towns on opposite sides of what was then a river curving through a deep valley. The forests—so often full of roving deer and bushes constellated with juicy berries—yielded no harvest. The river—so often full of flashing fish—yielded no catch. The people of the two towns, ordinarily peaceful, took up arms against one another. They ran raids, slashed open one another's bellies to fill their own, stole what they could find in one another's larders.

Mount Storm King watched with dismay—watched and watched until he could no longer bear the sight of his people turning against one another. At last, he pulled his great rocky head from his broad rocky shoulders. He heaved it down upon the people of the lake, shattering both villages, crushing the warriors where they stood locked in combat. Smoke rose from flattened fire pits. Water surged between the broken bodies and smashed basketry, backfilling where the fallen rocks blocked its passage downstream.

Survivors pulled themselves up out of the wreckage. They excavated injured people from under the debris, loaded them on rafts of splintered material and fled for dry ground. They understood that this was their god's way of speaking to them. They understood that the mountain demanded peace. And so they set aside their weapons, sopped the blood from their wounds, and went about the business of restoring what they had lost.

It's a true story, more or less. Geologists have confirmed it: nine thousand years ago, a massive landslide came tumbling down from the mountain and dammed the river.

Sea-run steelhead and cutthroat trout, blocked from returning to the sea, lingered in the lake. They milled about and hunted for whatever food they could find, and though the mandates of their bodies drove them to seek the sea, they circled quietly, learned to live in their new freshwater world. Lake Crescent, at the northern edge of Washington's Olympic Peninsula, is now home to two new species of trout—descendants of those trapped fish—that live nowhere else on earth. Life is pretty good at making the most of new circumstances.

The land, like the heart, is made up of layers of living material, layers of story stacked like beating muscle. I came to the Olympic Peninsula for a boy. We'd met on a schooner, where we both worked as deckhands in our final year of college. On hot days after hauling on lines, he'd fold me into damp, salty hugs that smelled like ocean. He wrote me songs on his guitar. One evening, as we went walking down a cobbly Northwest beach together, he picked up a flake of driftwood and whittled it into the shape of a heart. It was young love, exuberant and earnest and urgent. When the sailing season ended, we tried working apart, in separate states, but life felt sad without the other there to share it, and so I took a job in Olympic National Park so that I could live closer to him.

Weekends, he and I shared a loft apartment in the little city of Port Townsend. Our first night in our new home, bedless, we made love on a mattress under the tall windows and afterward lay listening to the tinkle of glasses in the wine bar below the loft. We watched starlight illuminate our empty new nest, and we watched each other, rumpled and naked atop our pile of sheets, and we wiggled with the loveliness of all that still lay ahead.

Life there was sweet. He wrote stories and I illustrated them. We let bread rise on the counter, yeasty and ballooning with slow warmth. He would sit by the window, looking out over the rainy courtyard below—rhododendrons heavy with bloom, a gray fountain with a woman's figure rising from it. In the rain-silvered Washington light that shone through, he wove nets out of seine twine with fingers firm from handling wood and rope—fingers that would later move across my body, fingers as heat-seeking and muscular as reptiles. *You can tell a lot about a person by looking at their hands*, he once said.

At weekends' end, I would drive the hour and a half back to Lake Crescent in Monday's still-dark dawn to stay onsite for the work week. He'd send me off with love letters, with breakfast treats, and—once—with a silver engagement ring, embedded with a disc of iridescent abalone shell. *I hate saying good-bye to you in the mornings*, he said. *I don't ever want to say good-bye to you again.*

On those drives away from him, the highway—like a collapsed

telescope protracting long again—propelled me into the world outside our little loft. I passed through the tight weave of fir and western red cedar forests. I pushed into the wall of rain that usually intercepted me as I left the rain shadow and entered the deep sog of the temperate rain forest. Outside Port Angeles, the road took a long, swooping turn as it crossed the Elwha River. Where the bridge was built, the river was more of a lake than a waterway, shadowed by cottonwoods and spread wide by a dam a couple of miles farther downstream. Past the river, I would straighten out of the turn and continue toward Lake Crescent, my new ring pressed cooly against the curve of the steering wheel.

Two dams plugged up the Elwha River—one below the highway, to the south, and the other above it, deeper upstream. They'd been engineered at the turn of the twentieth century to provide electricity for the growing town of Port Angeles. The dams were built improperly—there was a breakage, and a flood, and the leaks were shoddily stuffed with sunken old cars and cement to hold the straining river back. More importantly, they'd been built illegally, without fish ladders. Before the dams, ten species of fish journeyed up from the sea every year: bull trout, crimson-bellied Dolly Varden, sea-run cutthroats, steelhead that arrived by summer and ones that came by winter, and five species of salmon with indigenous Klallam names that swooshed and clicked on the tongue like water over rocks: kʷítšən, q̓ə́čqs, scə́qiʔ, ƛ̕xʷáy̓, hə́nən̓.

A treaty written up in 1885 acknowledges that the Klallam people have lived along the Elwha River for time immemorial. For thousands of years before colonizers arrived, Klallam communities buried their dead in boxes brushed with red ocher. They etched stones with careful lines—each carved rock a story held in the hand. They hunted and gathered, but mostly they fished: pushing elaborate riblike fish traps into the currents of small creeks, and building wooden platforms and weirs across broad stretches of the river, and paddling into the Strait of Juan de Fuca in cedar canoes hollowed with hot stones. With pronged hooks, and spears, and dark nets woven from nettle twine, they pulled up fat fish in a haze of steady coastal rain. Neighboring tribes would arrive for potlatches, where the hosts would give away everything they owned, because wealth wasn't about what a person could accumulate but what they could give.

In that 1885 treaty, the Klallam gave away much of their ancestral land in exchange for $60,000 and a three-thousand-acre slice of land, which, by the time I moved to the Olympic Peninsula, had been chipped away into only about a thousand remaining acres.

But one thing that the Elwha Klallam people never signed away were their fishing rights—though it wasn't until 1974 that a court decision reaffirmed this, allowing them to return to the banks of the river without fear of arrest. By then, both dams had blocked salmon passage for decades. The salmon that the Klallam had been promised in 1885 were gone, or mostly. The Klallam fought hard to bring them back. They did battle against the dams. They sent their children to college, to learn law. They played a long game. Decades later, as I was newly arriving in the rain forest, they won.

The year that I showed up to work as a field science educator in Olympic National Park, the largest dam-removal project in our nation's history was about to begin. My job in the park was, in part, to teach visiting middle school students about the history and future of the river and to guide them toward studying the ecosystem as it currently was, in order to build what scientists call a "baseline"—an inventory of how things are, to compare against what they will become.

Before I brought students to the river, I would bring them to the River Room. Indoors, we leaned over a lilliputian watershed:

winding canyons, lacquered banks, a bucket of tap water to kick-start the whole rain cycle in miniature. We squeezed little foam bricks into the tight space where canyon walls came together, and we spooned sand behind them, and we staged sticks in place of logs, and we filled the whole valley behind our barricade with water.

Then we engineered the removal of our dam. Sometimes, students would remove one brick at a time; sometimes they would punch the whole thing out in a single go. We watched the water cut down through the layers of backlogged sand, watched the sticks come jumbling through the narrow canyon. As we fed more water into the system, it sliced and braided through the sediment, lifting it and fanning it toward the edge of the model, beyond which, perhaps, waited an invisible sea. Which method keeps the water clearest, I would ask, which approach allows the trapped sediment to flush out slowly, so that the returning salmon can breathe?

It was all conjecture and approximation in that little room, of course. Outside the River Room, a hundred years of sediment, spooned in by glaciers and flushed down by rain, had gathered behind the two massive dams. When I brought students on a field trip to the lower dam, only half a dozen or so fish lingered in the lower pool, loitering, nudging against the leaky cement, looking for suitable substrate in which to lay their translucent red eggs. Four thousand of them would return this year to try to nest in the cobbly lower five miles of the river. Four thousand pared down from four *hundred* thousand. Five miles of river severed from seventy miles of salmon habitat that, a century ago, connected to the sea.

The scale of it all boggled. For our watershed model, we used a pitcher to collect our sediment and feed it back into the system. Outside the River Room, the sediment trapped behind the dams would have filled dump trucks parked end to end across the length of America, twice. Where the Elwha River flowed north into the Strait of Juan de Fuca, I took my students walking across a rocky beach, starved of fresh river sand for a century. Upstream, we took samples along the shores of the dammed lake, where a hundred years of trapped silt slurped at boots like quicksand. Once, I had to rescue a kid from where the thick muck tried to swallow him whole. Stand still for too long in that silken mud and a body could be lost.

Now, after a hundred years of fighting for treaty rights, the Lower Elwha Klallam Tribe had finally secured plans to demolish the dams, to drain the muck, to bring riffles and salmon back to the drowned places.

At last, balance would return. But, alongside the victory and anticipation of the dam removal, a secret sadness sometimes settled in my bones.

One winter morning, I drove through spitting snow toward Lake Crescent. It was a quiet drive this time of day except for the occasional thunder of a logging truck roaring in from the other direction. In a low glen just past the river crossing, my car hit black ice. It slid into the opposite lane, then back toward the right side of the road, back again into the oncoming lane and, at last, in a slow and unstoppable swerve, it crunched into the guardrail and came to a stop.

I sat shivering at the wheel—miraculously uninjured, miraculously unshattered by any passing logging trucks.

My partner drove out to rescue me, to tow the car back onto the road, to settle my nerves as the snow piled in drifts along the edges of the road and the banks of the river. I revisit that moment sometimes as if it were a photo in a scrapbook: a man walking on ice for me, a hazardous road ahead, the cold morning light spilling across my safe but shaking body.

It was a small moment in the story of that life I led, back when I lived in a little city near the sea with a tall, blue-eyed man. A small, careening moment, quick with adrenaline, softened by luck. It wasn't even particularly important—indelible, in hindsight, only because of the sharp panic, which helps to archive memories like quick Polaroid keepsakes. Four years before the crash, he and I went swimming together, mostly naked, in the groin-chilling water off the dock near the ship where we worked. We yelped and clambered from the sea, and though our bodies were stiff and stippled from cold, he gripped my arms and kissed me. A year after the crash, I'd leave our shared home, driving alone down another ice-slicked road—this time away from an engagement, away from an unfaithful partner, away from the rain forest and the life we'd shared there. There was a before, and an after, though sometimes sadness and time and the fickleness of memory blur the edges between them.

One afternoon, a friend and I slid a canoe across the mud and fallen cedar boughs and into the cool water that filled Lake Aldwell, above the lower dam. Our paddles sent audible drips plashing behind us in backward-trailing V's. Fleets of trumpeter swans rafted together in the distance, white wings luffing. We navigated up a quiet side canal. The autumn cottonwoods torched the water yellow, and what might have been a muskrat broke the reflection nearshore with a startled whiskery thrash. Under us, unseen, moved kokanee, native to this freshwater refuge for twenty lake-bound generations. I could see it: these glassy backwaters drained down to gravel, the swans scattered like confetti. I couldn't help but mourn the things that love lakes, the things that would soon be gone. The hydroelectric engineers were invaders a hundred years ago, surely, but were the swans?

In my wilderness medical trainings, instructors have begun to caution against "pulling traction" on a split femur—stretching the leg and splinting it so that the sharp bones puzzle piece back together again—unless we know that the splint will hold. A broken femur, startled loose from traction, risks damaging the bruised tissues and pounding arteries still straining to live at the periphery of the breakage. Here in the Elwha, engineers were preparing to crack open the dams, to pump thirty-three million cubic tons of sediment downhill. Would the salmon they were wishing to restore survive the turbid current? Would the splint hold? Or would loss be compounded by loss: muddy river, fishless, now swanless, too? Empty river, bled out?

In the final chapter of our relationship, I left for Alaska to work for a season in other rain forests at the mouths of other salmon streams. He departed for a summer in Maine, where he spent his days working on an old sailing ship and his nights falling in love with a curly-haired girl who was not me.

I was far away from the Elwha River when the dams came out. But when the removal process began, I could watch from afar. I leaned over my computer screen in a little café in Juneau, where a video ticked through all that I had missed. Swift time-lapse footage showed the barges notching the dam walls incrementally lower, the diversion of the riverbed, the bulldozers grooming and molding the ground just like the miniaturized canyons in the River Room. The sped-up footage rolled through snowfall and the smear of rain, showed autumn goldening and spring's murky floods.

And then, only a week after the last blast at the upper dam, I read online how Mel Elofson, who grew up a quarter mile from the lower dam, found a female Chinook swimming near the riverbank above the dam. His grandmother had walked the same shoreline as a young woman, he said. She'd watched the tall wall erected where the river bent between high cliffs. Despite time and trammeled treaty rights, despite sediment and dynamite, the salmon were already pushing back upstream.

When I think of that car crash, I think about simultaneity.

In the "then," he was the love of my life. The same moment, seen in hindsight, shows him for what he always was: a fleeting figure in the snow, a flickering time-lapse cameo in a life that would mend when he was gone.

Anyhow, this story isn't about us, not really—there are salmon in the sea whose lives are longer than our relationship was. In another dimension, maybe we never met. In another dimension, maybe we married after all. Maybe in a different dimension, the Olympic Peninsula stayed cloaked behind its veil of rain, unfindable, unloggable, uncolonizable, and the Klallam still fished for thronging salmon on the banks of an undammed, whitecapped river. Maybe in another world, the swans abided, and the dams held strong, and Mount Storm King watched over his cast of ever-evolving landlocked fish like an emperor counting his many coins. The point is, sometimes it feels like the worlds that were and the worlds that are to come and the worlds that never came to be all live close together: beating organs in the same body of time.

When I finally returned to the Elwha River, my engagement was over and the dams were gone. Salmon were continuing to flush upstream—sooner and in greater numbers that anyone had anticipated. Bulldozers had filled and flattened the canal we had paddled in our canoe. The shoreline was naked where former lake bottom had been groomed for seedlings. And rising out of the mud stood a quiet army of tree stumps. They were each as wide as I was tall, notched on one side where loggers had inserted springboards long before chainsaws ever came to the Olympic Peninsula.

These were the old giants—the forest that had lived before the dams, the forest that had been cut before the dams, too. They had abided under the dammed lake, a drowned stump city peopled by cutthroat and lit dimly by the bubble trails left by beavers. Exposed now, this ghost forest was finally free to collapse back into the clockwork of rot and weathering. The stumps would become soil; their roots would interlock with the lacy new roots of cedar saplings the way that a grandmother's hand might grasp a child's. A thousand years from now, an old forest might shadow the riffles of the Elwha again, might cool the water so that half a million salmon can shimmy across the polished gravel, can leave behind their luminous eggs.

I understood it, then—the way that a landscape, like the human body, is haunted by the things that were, by the things that might be. I could still see the swans. I imagined looking up through the water's refracted light at their pale bellies from where I stood now on the lake bottom. I thought about how, as we paddled across the dammed lake, the old riverbed had drifted unseen below us: a trough carved and re-carved and winding, silent as a ghost town, under the paddling birds, the fluttering alder boughs. Lake hadn't replaced river—not quite—and river, now, had not fully replaced lake. Instead it felt as if they existed atop each other, these layered histories of landscape. I thought of the time lapse I had watched while I was away—how a year of transformations had been compressed into a careening minute. I imagined how ten thousand years might also condense into a quick time-lapse revolution of gears: fish, forest, dam, lake, swans, dynamite, fish, fish, fish.

It reminded me of those phoropter machines that test vision, the ones where you press your temple into the cool metal of the machine and watch the world blur and focus through different lenses. One or two, the optometrist asks, two or three? And you choose, choose until the world collapses into focus, but the other lenses are all there, too, still nestled close against one another like dense cake layers, like worlds that could have been. Husband, or stranger? Or some third and intangible truth? The other worlds glint behind their casing, as real as unseen rocks under a river's surface. The Elwha felt like this to me as I watched the time lapse of the dam removal, and as I stood later among the logged giants, as I felt the glassy ghosts of past and future all around me.

Click back the lenses, watch them clatter out of sight behind their cool jacket of time. Settling into focus, before and again and for always, will come the fish, with names that clack like glass, with names that float like unstoppable water.

Mobulas

IN THE MORNINGS, THEY fly in the flat gold light. They are like gunfire, popcorn, hands clapped together. Our boat crawls along quietly, hours away from anchoring. The mountains are coral pink in the new day. The sea is windless and blue. *Thwack*, go the mobula rays. *Flutter, splosh*. They launch themselves out of the water—three feet, eight feet, ten. Airborne, their flat fins flap like wings. Their tails are whips. Like inverted divers, they emerge face first and come arcing skyward until gravity bends them, belly flopping, back down. Sometimes their balance shifts wildly midair. Their pale undercarriages flash, gills up, white skin shining, and they backflip and somersault and come colliding upside down against the surface. Divers call them and their kin "devil rays," but my friend from La Paz calls them *panqueques volandos*, flying pancakes, and it is infinitely more apt.

My work aboard the boat is to guide, to inform, and so people ask me *why*. Why do they jump? Why do they leave the water, their shining home, to gasp in the thin sky?

I'm expected to give answers. Maybe parasites, I tell them, shaken loose when they crash back against the flat surface.

Maybe joy, to go pancaking through the air. Do elasmobranchs dream sharky dreams of flying? Do their muscles sing with the strength of it, the freedom? I cannot know, but I can imagine it.

It could be momentum, I tell my clients. They swim so swiftly that the breakthrough could be accidental: zooming just under the surface and then airborne before they know it. They startle and swim about in the air, all out of place, until gravity smacks them back to the place they know.

A man in a BBC documentary explains it with casual confidence: Males fly to attract mates. The higher the flight, the splashier the landing. He gives this cocksure answer—but how is he to know what happens once the animal returns to its water world? Biologists have not tracked them, nor shown females thronging toward the biggest splashes. Females jump, too—and what's in it for them? He sells his tidy answer as truth, with no room for maybe. I hate him for it.

Aboard ship, I'll say good night to one lover and walk the quiet, yellow-lit outer deck to another man's room. The first is stocky, sweet-faced like an all-American baseball player, flushed with near sleep from a long day's work. The other is lanky—blue eyes like Alaskan flowers, his grip on my back like that of someone trying to hold on to something slippery and important. We do not kiss, but I let him hug me, and I'm not sure what I want or what he wants or whether we're the right people to give it to each other.

The first of these two will imagine with me what our children will look like. Soon, I'll leave him and follow the other across the globe. I won't know. Won't know what I desire, won't know what I'm doing even as I watch myself do it. We'll throw ourselves through cities and across mountains and leave each other abruptly. I'll do things to each of them, in turn, that feel cruel, necessary, tender, until we all three collapse out of each other's lives and my ears are left singing in the concussive blueness that remains.

One day, we go swimming off the boat, my clients and I. Near the rocky crumbs of islands where we snorkel, the water is blurred with bird shit. Huge, fat-lipped groupers slide under us like dinosaurs. Sea lions spin close and pesk me with their mouths. But out where the islands drop off into deeper water, things are quieter, and it is here—beyond the toothy sea lions—where the mobulas appear. They rise in a tight group and the white parts of their bodies gleam turquoise in the filtered light. I fill my lungs and swoop downward to better see them, but they are swift and wary and my dive herds them sideways and down. Their muscles seem to predict the movements of my own. They veer and sink and rise, pixelated by the distance between us.

Why is the question that we ask so often *why*, when *how* might yield more fascinating answers?

Why hunts for motivation, demands BBC certainty.

How speaks the language of muscle, of momentum.

How gestures toward pressure changes and gravity and the prickle of electricity that allows rays to identify where other living things are moving all around them.

How wonders at the implausible strength it takes to punch entirely out of one world and into another, even for a moment, flashing and flailing and suspended, motivated by nothing but the mechanics of a careening body—theirs, mine.

Below me, the rays move like a swarm of starlings, shifting and merging as one. Lungs aching, I rise toward the surface, where the unknowable rays grow bluer, grow distant, until they are indecipherable from the sea.

Witnesses

HIGH NEAR THE SURFACE: top smelt, with tiny forked teeth, with bodies swirling against silver bodies in a shifty murmuration. And among the swishing blades of seaweed: giant kelpfish, pulsing different colors like living mood rings. And cabezon—froggy-mouthed, fins like ribbed corduroy, eggs that are gelatinous, Bubble Wrap plump: a single bite can kill a man. Blue-banded gobies with highlighter-bright bodies, hanging upside down from rocky ledges like lanterns. Black rockfish and blue rockfish and olive rockfish and kelp rockfish. Garibaldi, cheddar orange. Morays, their unhinged snaggletooth mouths open in a grin, or in a grimace, though really they're just trying to breathe—and weren't you, too?

All these watched, as your diving companions surfaced. As you did not. As your body sifted through the fog of plankton, through rubbery curtains of kelp. Eyes among the seaweed, eyes among the rocks.

And when night arrived, bat rays winged over you, their sets of teeth generating, regenerating inside their mouths like endless necklaces of pearls. Close by in the kelp forest, horn sharks squeezed out dark, spiral-shaped egg cases that glowed under the strobing searchlights, the unborn bodies inside them backlit and pumping slowly in silhouette, the way that a heart pumps. Splotchy swell sharks with eyes like oil slicks glanced at you, glided past.

Your body knocked and settled in the deep current, ploughing patterns into the sand. Helicopters sent light swimming through the taut tops of waves. And when you'd been dragged up—waterlogged, unbreathing; when you and your gear had been cut open like fruit though no bad seeds could be found; when they brought you, stiff, back to our hometown crematorium in the brittle height of July, I could not bear to look.

Still, by the time the rescue divers found you, a thousand other eyes had witnessed your cool and emptying body. A lidless fish has no choice but to see, to see, to see.

A Fish in Winter

UNDER THE MISSION MOUNTAINS, under the casual surveillance of a white dog with black speckles, under dirt that was seafloor, once, in an eon ruled by drifting cells and early algae, we planted garlic in long rows. My friend's brother, a farmer, had riven the bulbs apart and staged them in baskets and buckets, which we scooped from as a bartender scoops ice. The men moved ahead of us, stamping the tilled November earth with holes. We followed behind, hunching over the beds. We guided each clove as deep as it would go, two fingers like lovers' fingers pressing the seeds three knuckles down into the Montana soil.

The seeds were white and blushed and papery and firm, streaked with purple like blood rising up gauze. As I tucked them into the earth, I thought about burial. I thought about the end of the latest work season that had brought my ship from Alaska back through Seattle, and had brought me back in contact with an ex-boyfriend. I remembered how we'd had beers, leaned closer to each other's familiar smell, returned back to his fishing boat to fuck on wool blankets. How he'd touched me with rough hands that felt tuned to some other woman's desires. How he'd forgotten what I liked, forgotten how to kiss the way we used to kiss, forgotten a condom, forgotten the work we'd already put into forgetting each other. I'd left lonelier than before. With each plunge into the mudstone till, my fingers worked toward some bigger act of burial, of disremembering.

I rose to stretch my back, and the dog pressed its cool nose into the scoop of my palm. I liked this work. I imagined myself migrating inland from the sea. I could see it: my own little garlic farm under some unshakable mountain range, my own dog at my side, a winter's waiting the work at hand. There was something soothing about the methodical movement, one seed at a time, each downward thrust a sort of blessing. I liked the idea of gardening after the autumn harvest, tucking seeds away at the threshold of winter like prayer stones. The garlic cloves needed the muffled quiet of deep places, the rattle of a hard freeze, no touching, no fussing. Come spring, they'd wonder if they'd been forgotten, gone so long out of sight, underfoot. They'd surface on their own time, according to some deeper clockwork of the earth.

Maybe it was too easy a metaphor: a fall burial, a spring rebirth. At the dusted tops of the Missions, along the brittling edges of rivers, winter was already encroaching.

A fish in winter buries itself where there's still some give—tucks itself low against the rocky hip bones of the riverbed and hopes, in whatever way fish might hope, that the ice won't reach all the way to the bottom. A fish in winter slows down, settles low. A fish in winter might recall, in its own fishy way, the juice of fresh mayflies crunched in spring pools, the refracted dance of summer light on water. Then again, it might not. Maybe it only knows the abiding northern dark, thinks neither backward nor forward, simply persists, unblinking, in the hollow belly of creek.

The garlic farm was a still point in an autumn of staggering movement. From Seattle, I'd driven to Montana to browse graduate programs, to shop around for a new life that belonged to me and not to the past. I spun through a few days with an old friend—mountain hikes, warm soup at her brother's farm—and then pulled southward on my own into Missoula, where I tried to imagine how life as a student might feel, nearly ten years after I'd taken my last college class.

I wanted a story that belonged to me. Without a sharp vision of what that might look like, I rose buoyantly toward new possibilities. My first night in town, a text message rolled in from a firefighter who'd picked me up, hitchhiking, the previous summer on an Alaskan road. We'd kept in touch, loosely. He, too, was passing through town, just for the night, on his way to visit family in what Alaskans call the Lower 48.

That summer, he'd passed me on the outskirts of Palmer, then circled back around the block to pick me up. It was a hitchhiking bonanza: a handsome driver, cheeks ruddy, jaw squared, plaid shirtsleeves rolled back across broad forearms that I hoped were not murderous. He drove me farther than he needed to and at the end of our drive, as I slung my bulging backpack down from the truck, he offered to take me dancing the next time I passed through. As he wheeled away, the glacial silt along the edge of the highway glittered and settled around me in the midday sun, and for a few long minutes while I waited for another ride, I wondered how a wildfire fighter might dance, how he might kiss.

I wondered it, still, the night I agreed to meet up with him in Missoula. We met on the university campus for an outdoor film festival, and his canvas jacket brushed against my arm and the light from mountain bikers onscreen filtered and strobed across his face. We wandered into the night parking lot and, *Let's get out of town*, he said.

We took his truck into the hills outside Missoula, stopped on a Forest Service road, and went crackling through the night woods in search of firewood. We built a fire right there in the middle of the road. He pulled out a fiddle and sent lonesome strains unfurling into the autumn air, and we lay back across his tailgate and shared stories about ourselves—his childhood in Alaskan foster homes, my dad's death—and as the fire died down, he dialed up the truck's radio and danced me, slow, under the full moon in the Montana woods. Just before sunrise, he put me behind the wheel and guided me back toward town, talking me through the machinations of stick shift. I stalled three times on the unpeopled dawn roads that merged through town. He smiled and waited and eventually, bumpily, we arrived.

See you never, I joked as I hopped out of the truck.

Come north to Alaska with me next month, he offered, and the too-perfect light of the setting moon caught his curls. Was this real, this dream following so closely on the tail of disappointment? *I've gotta get this truck home in time for Christmas. A couple old college friends are driving north with me and we could use the company.*

My next boat contract wasn't until the end of December. I needed to swing through California first, to visit my own family, to collect my winter guiding gear, but I could make it work.

Sure, I said.

It should have felt like a fairy tale, but the coming days in Missoula felt foggy and wrong. I walked to meetings on the university campus in spitting sleet and, uncharacteristically exhausted, I would drag myself back to the friend's studio where I was staying, sipping tall cups of coffee along the way, which steamed in the Montana evening like fumes from some careening little locomotive. My legs were heavy. My body felt tender, a clove rubbed raw.

On the long drive home across state lines, winter descended

along the steep highway passes, weighing conifers with November snow. I twisted in my seat, tired and tense with cramps that grappled in my gut like living animals. I guess I knew—before I saw the perpendicular blue lines on the pee test, before my hometown gynecologist confirmed it for me with a smeary ultrasound—that the things I'd wanted to bury from that Seattle visit wouldn't stay down where I'd pushed them.

In the summer bays where I guide, female salmon throw themselves out of the water again, again. Their bellies are full of eggs, plump and dense as cactus fruits. The fish smack against the water, itchy with the fullness of it all. Are they eager for their long journey upstream under the slow-moving shadows of trees and the quick-moving shadows of bears, where they will arrive at their bridal beds and die? Are they uncomfortable with the expanding kernels inside them? Do they leap for joy? Do they leap for practice? Do they leap to free themselves? To jostle that unfamiliar mass loose, as a sick mouse chews through its own skin to tug out the tumor lodged within?

In winter, a fish swaps agitation with stillness. A fish in winter eats what it can in the darkness, in the cold—nibbling dead stuff, idly scouring the riverbed for lost eggs. Days shrink short. Ice coagulates along the surfaces of streams.

I wanted it gone. My ex-boyfriend did, too. The answer was obvious, scarcely needed to be spoken, though I called him to let him know I was taking care of it. *I'm so sorry, hon*, he breathed into the phone, and I knew that he meant it. I knew, also, that he'd never try particularly hard to imagine that twisting feeling in my belly where an unwanted thing was multiplying—knew that he had the easy end of this arrangement, but that's how this kind of story goes, isn't it?

Twenty-four hours before the abortion, my gynecologist pressed a little sponge into my cervix, made from seaweed, meant to open things up, meant to make it easier.

The following day, he vacuumed me clean. It hurt. Thick blood sloshed into a glass receptable on the floor: an odd aquarium, swimming with mistakes, with sticky things that should never have been.

You'll want time to recover from this, my mother counseled. She, too, had had an abortion in the first year of her relationship with my dad, in a friend's lab in Boston in the early seventies, when women's ownership over their bodies was newly legal. She knew what lay ahead, though I could not. I nodded, tucked a bag of maxi-pads into my carry-on, checked into my northbound flight, and kept moving. The calendar ticked into December.

A fish in winter looks for warmth, when it can find it. Lake chub, with leaden scales that glint, with barbels like wispy mustaches, move closer to hot springs when winter encroaches. The hot springs stay liquid as the world around them freezes, as the night expands exponentially overhead. Escaped goldfish that have no particular right to be alive in the Yukon hunker down in warm mineral pools where comfortable water springs from the icebound earth. An animal likes comfort, whether or not it deserves it.

The firefighter and I convened again in Montana and, along with two college friends of his, had kitted his truck out with bright headlights and a sturdy moose bumper, like the cattle guards on old trains, in case of sudden wildlife on the dark road ahead. He bundled me in his oversize Carhartt jacket and

I handed him tools where he lay under his truck in the Missoula snow, tightening the last of the bolts.

Northbound, Montana farmland looked gray in the last light of the day: silvery stubble of dead growth punctuating the December snow. The Kootenay River followed the highway into Canada, or rather, the highway followed it, drawing close and pulling away again from the scrubby winter trees that marked where water flowed, unseen, in the gathering dusk. We crossed the border into night.

That first evening, the heat broke in the firefighter's camper. By the time we pulled over outside Cranbrook and discovered that the heater was out of fuel, the nearby stores were all closed up. And so we hunkered down: thick blankets, body heat. His friends shared a bed and, at the other end of the camper, he joined me, fully clothed, under lofted layers of sleeping bags. We fidgeted, looking for heat. He moved his body closer, slung an arm over me, and I softened into the scoop of his long body behind me. There was a brief moment where we hovered—touching, uncertain—before his fingers moved unambiguously under my striped long johns, across the bare skin of my newly emptied belly.

Was this mine to take? I was alive with want, and unsure what I was allowed, what I deserved in these early days after we had betrayed each other—my body and I. But here, in absolute quiet as the rest of the world dozed, as the camper rocked cooly in the wind off the Kootenay, I turned and worked my fingers under the firefighter's stiff Carhartt canvas, peeling away the bulky jacket, the thick pants tatty with wear. He kissed me and it was like the fires that I imagined him fighting—peaty, quick-moving, chemical. His breath heated my shoulders, my breasts. He took handfuls of me. Under the synthetic weft, his furred legs—mammalian, muscular—swam next to mine. It felt illicit, and warm.

The next day, the bleeding began. It came in waves—big, knotty clumps that flowed like spring breakup, dark and irregular. It didn't hurt, per se. But it felt shameful and alien and alarming: my body unburying its histories; some archaeological dig gone wrong. As we moved north toward the Yukon, towns came sparsely, quick intersections that appeared and faded miles behind on the unspooling road. I told the firefighter that I was on my period, unready for the vulnerability of telling the more problematic truth. When I could, I used bathroom gas stations, where I purchased tall Styrofoam mugs of hot tea and skidded on slick ice between semitrucks toward bathrooms with sticky doors and reels of thin toilet paper.

More often, when we pulled over on the side of the empty road, I crunched into the woods and, squatting, my skin pink in the steaming cold, I stained the snow red.

A fish in winter can't always find warmth. Fish cope using what their own cold-blooded bodies can provide. Silky slime to heal from jagged wounds. Polypeptides tucked in sluggish veins to keep the blood from freezing. When ice crystals begin to form inside the body, proteins, spiraled tight as fusilli, glom on to them. They crowd close, locking the ice up tight so that—though it strains to grow inside heart, inside ovary—it cannot. In this way, in winter, a fish moves through a frozen world with blood that stays—astonishingly—unfrozen.

We moved north through Banff, Prince George, Whitehorse. The road was desolate, rocked by the occasional thunder of a passing truck and then, for hours, quiet under the sprawling Yukon sky. Thin trees whirred outside the window and caught the low slant of winter light in their branches, like orange fruit. Days were short and the firefighter was interested in getting home, and so we spent nights under moonlight, alone on the road, rotating driving shifts, snoozing in the cab or in the camper when we could.

The moon was nearly full, and the countryside was crystalline. The land was a sleeping body; the frozen rivers, turquoise veins. The mountains shone with their own kind of light. Once, two moose big as mammoths crossed the road ahead of us, and one ghosted into the trees but the other skidded out on the ice, long legs spread in four directions. It locked eyes with us as, hearts in our mouths, we skated to a near stop and watched it wobble into the woods out of our collision path, rickety as a ladder. Lynx padded along the road's edge with paws like paddles. It was a wilderness of quiet.

I couldn't know, then, the journey that still lay ahead for me in the days after the bleeding stopped, in the months after my body tried to stitch itself together. I couldn't know how, when a fertilized egg is washed from the uterus, the body's estrogen production plummets and brain health goes haywire: lonesomeness, joy, despair. I couldn't know that I'd spend the coming spring browsing pharmacy aisles for something to make me feel hopeful again, that I'd spend the coming summer weeping at the edges of glacial rivers in the corner of Alaska where I worked, wondering how swiftly those waters might carry my body downstream, how quickly their cold might end me. I couldn't know that the wanting I felt in my rib cage, the craving I felt for the fireman's grip on my hips, was only the first in a carousel of surging and unmanageable emotions that would drag me through the coming year, deep and swift as rivers.

All I knew, in the slow nights that pulled us through the heart of the Yukon, was the pleasure of moonlight on snow. The warmth of broad hands on my breasts. The guilty terror of my own thick blood in the diamond-like darkness beyond the road.

A fish in winter is not like garlic. It does not thrive where it's entombed. It does not buckle upward into the slushy spring thaw like a phoenix. Lakes in the Yukon are cold and clear, as unmuddied and low in nutrients as the snow that feeds them. Under these conditions, plants and animals grow slowly. A lake trout might not reproduce until it's a decade old. In a government pamphlet printed in Whitehorse, a picture shows a trout suspended between two hands: gray body, crimson fins. It's over forty years old and only measures the length of a small forearm. For each long-wintered year of its life, it grew a single centimeter. In winter, fish don't dream big dreams, if the shape of a body is any measure for the dreams inside it. Fish hunker. To survive is enough.

I want to tie you to a tree, he whispered. *I want to cover you in honey, I want to lick you clean.* While his friends shared driving duty in the truck, he peeled me open in the camper, hinging my legs open, pressing his mouth into places that constricted me with bliss—again, again, again—as the Yukon road rattled beneath us. It was almost too much. I knuckled his damp curls, pushed away his flushed face, but he pushed back like fire, consuming me until all that was left of me was a shaking body of light. The aggression scared me and delighted me, equally.

One evening, when we were parked for a few hours of shared sleep, he and I bundled up and crunched together into the bright night. It was negative forty degrees out—the sort of cold that feels like small razors on open skin, the sort of cold that makes snow squeak like Styrofoam, the sort of cold that makes you cough when you breathe too deeply. Every mountain was a lantern, luminous with Yukon starlight. The firefighter led me down to the river. It had winterized into a firm and radiant road. We stepped gingerly onto the ice and moved downstream as commuters might pace a sidewalk. At a bend in the river, we paused to listen to the silence—the kind of thin-aired, drum-tight quiet that only the Yukon knows, in December, in the night. No audible water moved. The stunted spruce trees held their breath. Our own breath, hot, froze like ghosts.

As if compelled by some unheard music, the firefighter took my hand and spun me on the tractionless ice. Just as he had done in the Montana woods, he danced me. With killing cold encroaching through my puffy layers, with my exposed hair crackling like spun candy, with river ice creaking where we twirled—it felt both dreamy and dangerous. Underfoot, fish swayed beneath their ceiling of ice—fish with empty bellies, fish with eyes like moons.

A fish is cold-blooded: like seductresses, like killers. How silly is it that an animal with a heart that beats so differently than ours got tangled up, somehow, in human value judgements? What does a cold heart know that a hot one might not? A fish in winter doesn't waste energy working to be hotter than the current around it. A fish is poikilothermic, from Greek *thermos*, heat, and *poikilos*: varied, painted, dappled like light across a streambed. Its body temperature does not stay stable but swings and falls as the world around it changes. If a fish worked to stay warm, the water around it would pull its heat away as steadily as a vacuum cleaner. In the Yukon, under ice, a fish's body is cool as ice.

When I was small, I'd have recurring dreams of violent things enacted on me—witches slicing me open in underground kitchens, dinosaurs chomping me alive. I'd wake from the dreams feeling real pain knifing through me. Sometimes I'd only wake halfway, in a state of sleep paralysis: pressed heavily into my bed by something invisible, unable to lift my limbs or to call out. In the periphery of my room, where I could not turn my head, something dangerous always seemed to move—a thing that was not frozen like me, a thing that might do me harm. This is how I woke in the back seat of the truck one night—to voices moving over me in the dark as the firefighter and his two friends worked their way through a midnight conversation about women's bodies.

I just don't think abortion should be legal, he was saying. *If two adults are having sex, they should accept the responsibility of a baby.*

What about rape, or something like that? one friend asked.

Well, I think rape should be punishable by death, the firefighter replied. *Maybe a girl who gets an abortion should get the death penalty, too.*

His words were like winter.

In the dark, I shrank. Already, I was dizzy from the messy finality of my last relationship, from the raw physical facts of the abortion, from the intoxication of this new romance, from the unease and bewilderment of trying to balance my body's shame with my body's desire, from this careening and dreamlike road leading north. I knew that I had made imperfect choices. Vacuuming the fertilized egg out of myself had hurt—would continue to hurt in unexpected ways for months to come—but nothing about the procedure had felt careless or murderous. It had been inconvenient and necessary.

The conversation in the front seat felt flammable, violent. I found myself wondering—as I had wondered about kissing, about dancing—how a wildfire fighter might do damage. With his words, in that inescapable truck in the Canadian night?

With his hands, rough and punishing, when I told him why I bled? The backseat window against my cheek was cold as surgical steel; the land beyond it lonely with light from distant stars.

In deep dreams, we become poikilothermic: cold-blooded, dappled like a riverbed. Blood pressure and heart rate slow. Our breath becomes irregular. Studies in neurobiology describe REM poikilothermy as a state in which we become vulnerable to external hazards. We lose the ability to moderate our temperature. Unblanketed, we might freeze. Our muscles lose their power to contract, making us slow to startle and slow to react to threats. Dreams—even the loveliest of them—lock us inside our own kind of winter.

As a child, when I awoke into sleep paralysis, unable to shout or shake myself, my only option was to descend uneasily back into sleep, into those dark waters where I might escape into some less dangerous dream. And so, in the back of the truck, I did not speak. I curled up, clenched and silent. I waited for the risk to pass.

A fish in winter doesn't always survive. Imagine it: a block of ice cut free, squared at the edges, a fish suspended inside it like a scorpion folded in a novelty shop lollipop. It looks more airborne than icebound—adrift amid a galaxy of crackled light, gauzy as asbestos. The fish is a statue of itself, as stony as if caught in the gaze of some Medusa. Steely fins, firm belly. Its death is thorough: deeper than water, deeper than dream. It is not encroaching ice that killed it but instead the water below the ice, too frigid to hold oxygen. The fish suffocated in the airless current and, like a strange rock, floated upward, where the ice memorialized its death as neatly as a taxidermist might. Winter kills in unexpected ways.

Those nights driving north, while I watched winter bloom beyond the windshield, I hadn't reckoned that winter was growing inside me, too. Hollow womb, skittering storms of memory and desire and pleasure and guilt. My mother had warned me to slow down, and instead, like one of the semis ahead of us churning snow into glittering flurries, I had barreled northward, so eager for warmth after a season of disappointment. There, then, in the back of the truck, it all came slowly to a stop. I leaned against the door, silent as a trout. I settled into myself, felt my cool edges where they dropped off on all sides, no land in sight. Loneliness rang in my ears. And just as real as the winter inside, I felt it around me, too: four strangers in the cab; islands, all of us. The longest night of the year pooled around me like water.

Northern fish are descendants of the cold. For three million years—years of deep freeze, years that reeled across the Yukon in grinding braids of glaciers—enormous lakes formed and drained. Glacial melt tore valleys apart in cataclysmic floods. Fish were stripped away in the outwash. The ones that lived were sly, or lucky, or both: sequestered in waterbodies under the highest mountains, which still jutted like islands above the ocean-like continental ice sheet, or surviving in the scattered sweep of Beringia, that ice-free land that connected the North American and Asian continents.

When the deep ice retreated, these populations of fish stayed separate, relegated to their familiar lakes or locked inside a single river's course like those deep neural pathways that wed us to our habits. Some populations took territory, spread into new waterways. But on color-coded population charts, many communities of fish still linger in tight patches and bands that turn maps of the Yukon into globby rainbowed collages—living legacies of global winter. It is a landscape of loneliness. Watersheds unroll side by side, separate worlds of life within them, untogether, untouching.

I peeled away from the truck, the road, the winter as soon as I could. We parted ways in Anchorage, where I hugged the firefighter in the airport lobby, so bright with noise. Layers of armor padded our embrace: his canvas jacket against my chest, the blanket of loneliness that had draped around me in the preceding days, the muffled insulation of secrets that divided us.

From Alaska, I caught my flight to work and watched Hawai'i's archipelago unfold below, one lush island beyond the next, false summer glittering between. A thousand miles off, the garlic bulbs I'd planted still hunkered, half forgotten, under the Montana dirt—from planting till bloom, they'd take a perfect nine months to grow. Or so a friend told me; I'd be worlds away come summer.

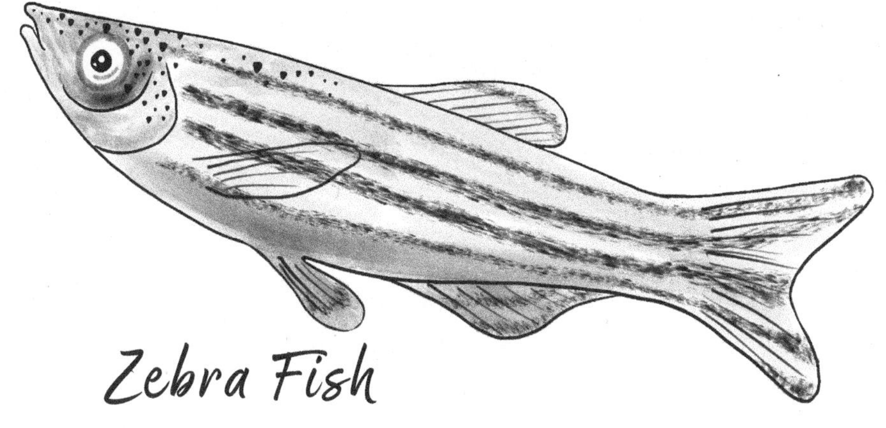

Zebra Fish

IN A WINDOWLESS LAB somewhere, a scientist feeds hydrocodone to fish. He begins with food—giving the fish the tools to feed themselves: a motion-sensor platform, a thin metal tube leading down into the water. When a fish swims over the platform, a green light flashes and a little snack travels down through the tube. The zebra fish mill slowly in their white tank, visiting the food platform occasionally, then drifting away to nudge at the edges of their home, to flit around like little bicycles coasting: quick kick, long glide.

Then the scientist changes the input. Instead of food, drugs come through the tube when a fish triggers the sensor. It takes them a few days to learn the trick: that swimming over the platform will make them feel good. That they'll stop feeling pain. That their quick little bodies, instead of floating, will fly.

Within a week, the fish crowd close around the platform, zooming and rioting, releasing more and more of the drug. The green activator light strobes above them. The opioids—like the little metallic balls that settled into holes in those handheld childhood puzzles—lock into receptors in the fish's brains.

They plug up the hurt. They send those little finned bodies sailing with euphoria.

The scientist fidgets with the controls. He begins to raise the sensor platform higher, making the water shallow and shallower still, knowing that the fish fear shoals, knowing that they avoid danger at most any cost.

Instead, driven frantic by craving, the fish swarm the ever-shallower water around the platform. They get reckless, aggressive. Their bodies and their brains—70 percent identical to ours—clamor for bliss in spite of risk. The fish cannot get enough. They buzz across the platform with their lean, striped bodies, flashing their long, moon-blue fins. When the drug gets taken away from them, they grow listless, anxious.

The scientist leans over the tank, rubbing his stubble, rumpled after weeks of this. He wonders what the next step is, wonders how to pull the fish away from the sensor, wonders what it might take to loosen the drug's hold. He tugs at his sleeves. Maybe he has a sister—as I have a brother—whose brain, like the zebra fish's brain, craves that bitter swallow, that floating high. Maybe, like my brother, she nurses needles between her toes, fades into some kind of glassy-eyed faraway place. Maybe she, too, cashed in on her retirement, sold her home, drew out on credit card after credit card to buy enough to fly for just a little while longer. Maybe she also dragged herself, uninsured, into a hospital one night after getting thrashed bloody in a deal gone wrong. Maybe she, too, had always been careful, tidy, kind. Maybe her art still hangs on the walls of her parent's home. Maybe, now, she breaks things. Maybe, now, she tells her mother she hates her, twists her mother's wrist in the driveway till she shouts with pain. Maybe her addiction is too big a place to be lost inside of. Maybe she's unfindable, in there.

The scientist stands up. The agitated fish have made his own heart flutter, quick and anxious as fins. He rises to go home. He rustles his answerless papers into a pile. He sets the computers to sleep, clicks off the overhead lights. He stands in the doorway for a moment, looking back. Above the fish tank, in the cave-like darkness of the lab, the scientist watches the green light clicking on, on, on.

Love and Other Fish

A SURVEY

A FISH IN THE hand is like a half-remembered word on the tongue: buttery, evasive. It edges away from your scooping palm, feeling you before you arrive, sliding sidelong like a gull with a cracker, not wishing to draw attention to itself. When your fingers close around it, it still somehow slips loose, like one of those ungraspable toy store tubes filled with water and beads. It's a wriggle, technically, but it feels more ghostly: gone before you even feel the fight.

Finally, you manage to lift it above the water, above the net, squeezing it gently between your palm's upper edge and the thick of your thumb. It's maybe three inches long, and you can just make out the hairlike indentation that runs lengthwise along its side—the lateral line that gives the fish its sixth sense, that lets it feel the water displaced by your hand long before skin meets scale, that allows it to follow its prey's vortices through the cool shadows under the bank.

Under its skin, the fish has a thin band of red muscle. This powers its quotidian swimming. But wrapped around its core are fat slabs of white muscle: the emergency gears. In a moment, the entire fish can galvanize, one thick fist of muscle punching outward and away. The fish's body is mostly made out of potential energy: quiet white muscle waiting for disaster. This is what lets it slip through your fingers like jelly. This is what keeps it from being netted at all. Around you, among the pebbles under your soles, in the deeper cutaways where your net couldn't scrape, are a dozen other fish, muscles thrumming, uncatchable.

How I ended up in a canyon four hours north of Tucson with fish between my fingers has equal parts to do with curiosity and disbelief. I moved to Tucson in the late summer, after the monsoon rains had mostly subsided but months before the desert heat would abate. The sun felt vertical and aggressive. The plants—cactus, yucca with sawteeth—drew blood. Weird bugs gathered around my porch light and the cracked pavement radiated heat even at night. Rivers that were illustrated blue on Google Maps were gravelly and dry when I visited them: washes that were just memories of water that might have flowed there decades ago, in a milder climate.

One evening, sipping beer with a friend in her living room west of campus, I saw the paintings: watercolor fish swimming on the wall above the couch. I asked her about them. *My boyfriend painted them*, she said. *He studies fish here.*

Here in Tucson? I asked, incredulous.

Yeah, she said. *In fact, I'm pretty sure there are endangered fish living right here in the city.*

I looked it up when I got home. Sure enough, there it was, in the headlines from 2016: "Endangered Fish Returns to Santa Cruz River Near Tucson." I'd seen Gila topminnows, little silvery captives hovering in their tank at Tucson's Desert Museum. The last of their kind, I'd thought. Vestiges of a wetter time long gone. Of course, I knew that there were fish still swimming in wilder parts of Arizona. But I tried to overlay this knowledge across what I knew of Tucson—its flat sprawl, its dry streets glittering with broken glass, its hot and empty riverbeds. I was captivated by the story, by its unlikeliness, by the hope embedded in it: "Hostile Place Supports Life."

That first year in the desert, I began going on walks with a man who studied soil and rigged monitoring equipment for a science facility north of town. On our early dates, we hiked partway up paths and then abandoned them, sliding off-trail among cacti and scree in search of private waterfalls. At night, we rollerbladed along the paved route that connected Tucson's dry washes: he instructing me how to fall gracefully, I resisting, eager to stay upright. When I gave up on blading lessons, we'd sprawl on the hot cement, facing the constellations, talking idly and listening to our wheeled feet tick and spin in the evening wind.

His legs were the legs of a rollerblade champion: long and tightly muscled and taut with dancer's rhythm. His kisses were the soft and thirsty explorations of a wild animal.

When I spent the night in his bed, he'd wake up before me and sneak off to work at dawn, leaving me to bathe in the slow morning light that arrived through the south-facing window. He'd wake me from afar, programming music from his office to play through his bedroom smart speaker. I was envious of Alexa—how she responded so swiftly to his requests, how she listened in on our secret conversations—but in the mornings she'd play me funk and disco and dance me out of bed, and I couldn't hate her too much.

He trusted me to let myself out in the mornings. I'd lock up the outside door, would shimmy open the chicken coop where the keys were kept. I'd move through the veil of desert dust and wing fluff, would sidestep past the birds that appraised me with dinosauric interest. I'd sling the key ring onto their rafter nail and step out into the day smelling softly of sex and of warm feathers.

Fascinated by the return of the Gila topminnow to Tucson's waterways, I wrote to Peter Reinthal, an ichthyologist at the University of Arizona. I wanted to see his records. Together, we scrolled through old sightings and surveys. The first Gila topminnow was identified in the late 1800s along the banks of the Santa Cruz River here in Tucson. The last one was collected in the 1940s. Seventy years went by before the little topminnow—assumed to be gone from Tucson forever—returned.

It's what we do, Peter explained to me as we pulled down jars of fish from the shelves of the university's collection. *We run long-term surveys so that we can keep track of changes in population.* Things might disappear, he explained to me, but they also might return. He handed me a jar from decades ago, it's label yellowy and etched in tiny print, little fish curled inside like bookmarks.

Peter conducts Arizona's longest continuously running fish survey. Twice a year, he brings a handful of undergraduate students to Aravaipa Canyon and, in concert with the US Forest Service and the Nature Conservancy, wades down eleven miles of the creek in search of native fish. He invited me to join his team, and although it wasn't Tucson—my questions about urban fish left floating for a later time—I tagged along.

When I joined him, the canyon was just waking up. Birds said their morning things from somewhere inside the green tangle of trees that lined the creek. I rattled down a dirt road alongside seven biology students in the back of a once-white pickup truck, ducking to avoid trailing vines. When we hopped out of the truck, a hazy glitter puffed up from our footfall: silt and mica and other dusts rubbed loose by moving water.

On either side, the canyon walls—pink, bruise-dark yellow—gathered close like escorts. We stepped single file into the creek. Cold water oozed inside our neoprene socks and warmed slowly as we walked. We moved quietly, although we didn't need to.

DESERT SUCKER

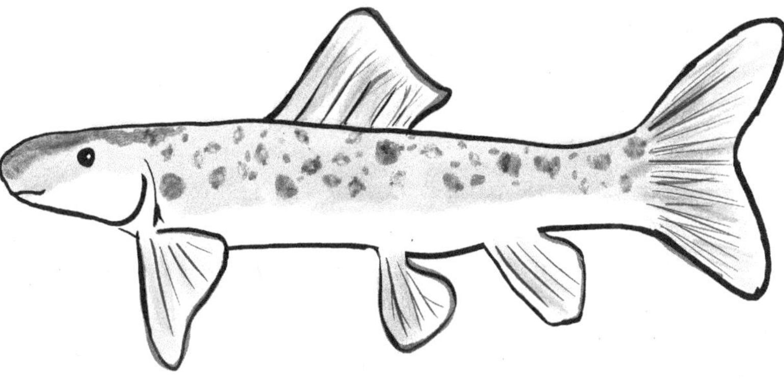

SPECIES	*Catostomus clarkii*
NUMBER OF INDIVIDUALS, SITES 1–9	198
DESCRIPTION	We hold them with their backs against our palms in order to see the shape of their mouths: perfect gasping O's for Sonora suckers, angular D's for desert suckers. The adults have bodies as long as my forearm. They cruise along the edges of our buckets, heads dinosauric and armored.
FIELD NOTES	They're the biggest of the native species that we find. I stand with a heavy bucket in hand and, from the edge of the stream, I can see in advance when a sucker will end up in the moving net. Fat shadows torpedo their way ahead of the lead line. Sometimes, they manage to slip over the edge before the seiners can lift their net. With suckers, we can anticipate what we might catch, and can also watch it get away.

The rollerblader and I would run out of words on our walks and it was OK: a comfortable hush that let us listen to the world rustling outside ourselves. At the county fair, we each ordered foot-long corn dogs and tried to keep them down on fast-moving rides, and at the top of the Ferris wheel, I knew that I wanted to keep kissing him for a long time.

It was easy, month melting into month. There was a gravity that made me want to be close to him, head pressed into the warm place between chest and shoulder, and the quietness was intoxicating, too. We never talked about the past, never talked about hearts. I was in love with the unfussiness of not having to name anything out loud. One afternoon, as we were preparing to leave for a hike, I told him so.

I would have been scared away if you'd tried to put a name to what we are too soon, I said. But the air shifted when I said so. I could see the tension in his jaw, where unseen teeth clenched beneath cheek.

It took him hours to respond. We drove to the top of our mountain, hiking for quiet miles among hoodoos and desert pines. We stopped to watch the sunset, picked our way the final few miles in the dark. We hadn't brought head lamps, and we walked carefully over the ice-crusted ground, letting the starlight illuminate the snowy path ahead. An owl called from the place where trees grew darkly along the creek bed. We listened to our breath, listened to our heartbeats, listened to the quiet, and it was clear that the quiet was no longer quite comfortable, that the stillness bore risk.

Finally, as I drove us back into the glittering sprawl of Tucson in the dark, he answered. *I don't like it*, he said, *not naming it. I know what I want. I need you to know it, too.*

The lights along Campbell—the locked-up coffee shops, the late-night taquerias, the Trader Joe's—flashed by in the side-view mirrors. His long fingers worried against each other in his lap.

I hedged, kicking away from this heavy thing that he wanted.

I need some time to think about it, I said, and I bought a plane ticket, and I left without giving him an answer.

A survey isn't easy work. In Aravaipa, two people handled a net connecting two wooden staffs. They tilted it and plunged it forward as they splashed downstream on either side of the creek, stretching the net perpendicular across the width of the water. Lead weights at the bottom of the net kept its leading edge close to the streambed, flushing fish upward into the weave. The end of a run took skill: Leave the bottom of the net too long on the bottom and the fish will continue downstream. One netter ran ahead, brought the net parallel to the bank, and together the netters stepped back from each other, pulled the material taut.

Bucket! a netter shouted, and surveyors came with sloshing buckets to scoop fish up with their hands and into their pails. As they identified the fish in the net, they shouted to Peter, who kept a tally on a weatherproof clipboard.

One juvenile desert sucker.

Two spike.

Loach minnow!

My first time handling the net, my side of the creek dipped into deep water, and I fell up to my shoulders. *Keep going*, someone shouted through the chaos of water, and I scrambled to my feet and kept the net moving forward.

In places that were too rocky for an easy scrape of the bottom, the netters set up the seine downstream. Kickers lined up several yards above and, splashing and slipping, stomped their way toward the net, hoping to herd fish with a frenzy of feet.

Peter stood at the creek's edge, choreographing the affair with a tone midway between urgency and preemptive disappointment: *Lead line down, lead line down, turn! Stoop. Pivot.* If a fish gets away, it's our fault. We try hard to please. We fail a lot. Fish get loose.

SPIKEDACE

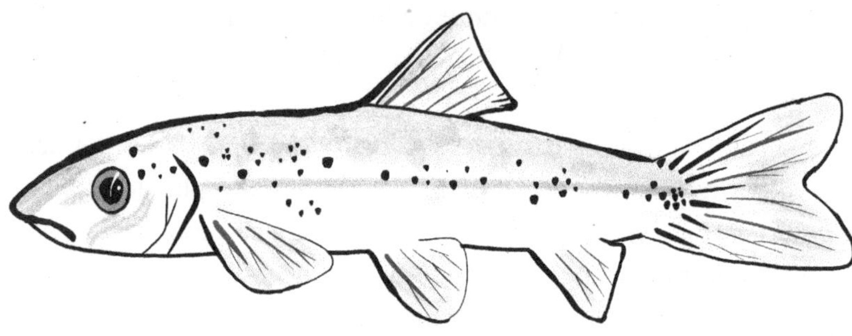

SPECIES — *Meda fulgida*

NUMBER OF INDIVIDUALS, SITES 1–9 — 192

DESCRIPTION

Imagine a handheld fan made of tissue paper: diaphanous material webbed between straight staves. These are the fins of the spikedace—gossamer skin pulled between fin rays, which, like fans, like finger bones, can spread wide and fold closed. A spikedace has one strong, sharp ray on the leading edge of its dorsal fin. I chase the fish around the net, where they wiggle and bend, grasp them with my fingers closed around my palm. I use my thumb to feel for their spike.

FIELD NOTES

Does risk make us love a thing more? Our buckets fill with spikedace: 17 at one site, 42 at another, 76 in one exceptionally spikedacey stretch of the stream. Spikedace! Spikedace, spikedace, call the netters. I grow to like them. They are abundant, recognizable, easy to tally. I do not learn until later in the day that most spikedace in Arizona have disappeared, maybe more than 80 percent of them. Where we are today is among the final streams where spikedace still haunt the riffles. Arizona is inhabited by the ghosts of waterways that once were, is haunted by the shadows of fish that are going quickly, almost gone.

When I was little, I walked out onto our upstairs porch with my mom. I could smell the oaks and the weedy backyard down below, the summery scent of warm chlorophyll and dust. All our windows were open to let in the evening air, and sheer curtains clung to the screens as the house that my dad had built breathed and settled.

When Papa asked you to marry him, did you have to think about it, or did you know? I asked.

I said yes right away, my mom replied. *Your papa and I had been in love for a long time.*

I nodded, my strawberry-print flannel pajamas and the cricket-loud air and my mother's answer all feeling soft and perfect.

I don't want to have to think about it, either, I said, leaning against the cool rail of the porch. *I want to know.*

Surveys are one of our earliest ways of knowing. That's what a survey is, after all: a way of seeing, of knowing the shape or number of a thing.

In Egypt, when the Nile overflowed its banks, ancient geometers waded out with ropes to assess the land that had been flooded and to restore boundaries between farms. In these earliest recorded land surveys, they moved through the shallow water, dizzy fish knocking against their shins, mud mucking between their toes. They tied knots, measured the shape of things, restored order.

A "survey" class at a university gives introductory information about a topic to a general audience of students: Survey of the Human Body, Survey of American Poetry. This belies the depth a survey allows, though: By definition, a survey isn't a glancing overview but a thorough look, a careful one.

In the northeast corner of Kansas, a graded gravel trail leads into the heart of a protected corner of land where a man named Henry Fitch followed behind copperheads and knelt to watch blue-tailed skinks. For fifty years, he took inventory, scribbled notes on behavior. His was the longest herpetological survey in the world and his little slice of Kansas might be one of the best-known square miles on the planet.

Every winter since 1900, all across the Americas, tens of thousands of volunteers step outside into jungle-loud dawns or sidelong snow to count birds. With each annual inventory, the Audubon Society Christmas Bird Count can provide a sharper picture of how bird populations across the continent have changed in time and space over the past hundred years.

A survey settles in for a patient view. Its perspective shifts and sharpens with each layer of information: *This is how things were. This is how they are now.*

Recently, Harvard released the results of an eighty-year survey that documented human happiness. It wasn't an experiment—subjects went about their daily lives and answered questions.

Are you sick?

Are you lonely?

Are you in love?

Each time the subjects sat down with their questions, they must have needed to run their own version of a survey, too, inventorying their insides from tip to toe.

Do I hurt?

Is this happiness?

Am I capable of love?

At the end of the day, we still understand very little about our insides. A recent survey conducted in Lancaster, England, showed how few of us can visualize what's going on under our skin, and where. *Where are your adrenal glands, your lungs, your heart?* The only organ that 100 percent of the subject pool could place correctly was the brain.

It took me a long time to take credit for my feelings, growing up. In perfect moments, the oaks and the poppies and the sprawling county park uphill leaked their quiet into our Northern

California house, but family filled it back up again with noise. My dad worked far from home for weeks on end—on ships and in ports doing engineering work. He was great at loving us, my older brother and me, but not as good at disciplining us. Instead, alone, my mom fought to build boundaries and consequences for a brother who, lost in a system that didn't understand him, spent his high school years in rotating, often rageful relationships with alcohol, pot, cocaine. He crashed his car one night on the oaky hill above town, spent another night locked out of the house pounding ferociously on windows, wandered into my bedroom sometimes in the early hours of the morning with a flashlight, scared and looking for company, and I would curl up and pretend to be asleep because I didn't understand.

He and my mother did battle. Their shouts rattled the house.

Unnoticed, I'd step outside. I wasn't sad or scared or mad, as far as I could tell. I got good at emptying myself. I'd sit in the swing that my dad had rigged in the backyard, and I'd watch the squirrels and let the noise turn to background static. I'd rock until the evening light settled around me, until the frayed rope impressed into my palms.

When the house got quiet, I would curl next to my mom in the bed that she shared with my dad when he was home. *You're such a good listener*, she would tell me, and she'd hold my hand, and we would sleep.

As a girl, I was a streambed. Things washed over me: dapple and light. Can a river know what lives inside it?

I flew across the country to go to college. *Aren't you going to miss your family?* someone asked, and I didn't know for sure, only knew that the distance—the expansive space between coasts, the promise of quiet—excited me the way that thermals draw hawks. It wasn't until those university years that I began to give myself permission to name things that I was feeling. I knew myself, I thought. Buoyant, happy. A good listener. But as I measured the shape of my new world and my place in it, I started to learn the names for other things, too: disappointment, jealousy, unease. Had these always been there?

What does love feel like? I asked a college friend once as I navigated one of my first relationships. *How will I know?*

LOACH MINNOW

SPECIES	*Rhinichthys cobitis*
NUMBER OF INDIVIDUALS, SITES 1–9	36
DESCRIPTION	During the quiet season, loach minnows look like the gravel on the streambed, like the stones brindled with algae, like cool shadow and a little bit of light. They keep close to the cobble. Their air bladders are too small to lift them into the drift to hunt for food, and so they snack on larval things that like low places. When they're ready to mate, though, their bellies turn to fire. Their fins, too: dazzling flashes of safety-cone orange. In the riffles, among the rocks, they are like torches.
FIELD NOTES	The ones we catch are sick. Dark fungus bleeds through the orange on their bellies, their backs. It's a new disease—Peter doesn't know it—but every loach minnow that we net has it. Wading near the edge of the stream, I find one torn open in the shallows. Its eyes bulge in death. Did we stomp it, or did it die of fungus, bloat? Maybe in our hurry to catch a thing, we had been hurtful.

After my dad drowned in a scuba accident, I was in charge of sifting through his things. I came back from college and inventoried his papers, his artifacts from a lifetime of shipping out to faraway places, collections of natural history clippings, photographs of trains, of our family, of my mother laughing next to a swing in a New England backyard.

A survey of his closet pulled different results: in a box on the highest shelf, I found a tin full of beautiful little love letters. They weren't from my mother—they were all written in the last ten years, by a married librarian living two states away in the city where my dad had been assigned to work as a port supervisor. They were lovely, almost girlish, scrawled in calligraphy. Some were poetic, some were raunchy. I cast around for what to feel about my father, whom I had always adored uncomplicatedly. Disappointment? Gladness that he was not wretchedly, wholly alone all those long months apart from his family? Discouragement?

An infidelity survey conducted last year showed that 36 percent of married men will be unfaithful.

My mother found the letters eventually, too, and still loved him, grieved him. *He was lonely*, she said. *He always loved his family more than anything.*

I couldn't have predicted that, a few years later, I would be pulling down my own box of love letters from the top shelf of my childhood room. How the man who'd sent them to me made me feel confident about love, made me feel confident about how the future would look. How the man who'd sent them had proposed to me one morning with a shimmering abalone ring. How he had cheated on me, lied about it. How I'd spent that New Year's Eve listening to fireworks as I packed bags into the back seat of my car and drove away from the only thing I thought I could ever love.

I considered burning the boxful of notes. My whole body swam with things for which I had no names.

RED SHINER

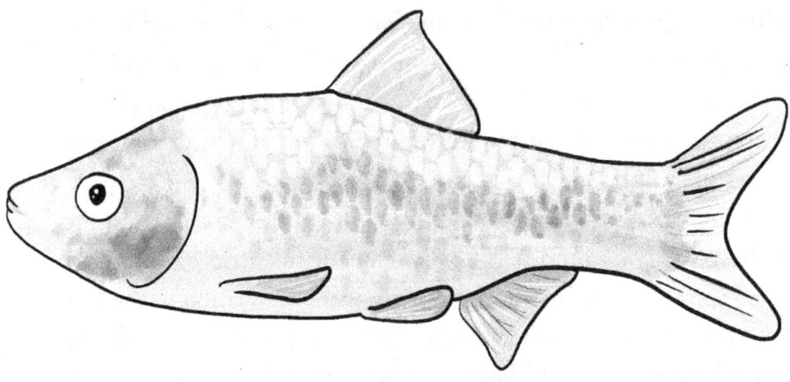

SPECIES — *Cyprinella lutrensis*

NUMBER OF INDIVIDUALS, SITES 1–9 — 1

DESCRIPTION — We catch one—a female—and her body is small and silvery blue and vertically flattened like the leather of a toe pad squeezed too long into shoes too narrow. Her carpy mouth gapes like a goldfish. She will release her eggs in batches of 70, 16 batches in a day. She eats eggs, too: those laid by spikedace, loach minnows, suckers. She munches up their pearly flesh, their translucent goo, vacuuming the stream floor clean. She is out of her range here, and thriving.

FIELD NOTES — Kill it, says Peter, but she white-muscles her way loose and is lost. We stomp blindly after her, flushing the gravel bed, churning the water into a D-day roil. We try to crush her. We've found a thing that we don't want. Netted and loosed and maybe still flashing somewhere in the deep places we cannot reach, she dirties the survey. Is one red shiner significant? Will we find none next year, or many?

What happens when you find what you don't want in a survey? After my engagement, I went netting for love long before I could name what I was pulling up. In Nepal, toes gripping the slate floor, bent in front of a lover, I gasped with what could have been pleasure—I wanted him to think so—but I considered, curled naked against him afterward, listening to Kathmandu's flower vendors in the street below, how he never touched me with his mouth.

It is what it is, he'd said when I told him I'd miss him, and so I came to approach men sidelong, without expectation, trusting that they only wanted small things, thinking that I could give it to them and plan my escape before they had a chance to disappoint. It got easy: kissing without love. Behind a guard station on an empty marina boulevard in Mexico. On the floor of a friend's bathroom with a man that I already hated although he hadn't yet given me a reason. In a hotel room in Seattle, on a beach outside La Paz, in a ship's cabin off the coast of Maui.

A man that played beautiful music on his guitar pulled me tipsily toward a half-finished cabin in Alaska's late-night dusk, and we fucked on the ground, his greasy Carhartts scrunched at his knees, wool hat knocked off.

This will have to be our secret, he said between thrusts, dim sky shining through wood framing. *I'm dating someone right now. We'll have to meet here, on the back road. We'll have to be quiet about it.*

I pushed away violently, rolled free of him, buzzing with something unnameable. *God, I didn't know. I have to go*. And I found my way back out of the thin boreal forest, sharp carpet of lichen crunching underfoot, stars watching me coldly.

Whatever I was pulling up, again and again, was toothy and unkind. My twenties were the color of white muscle, were the thirst and confusion of skin wanting skin, were the panic of pushing away before it could hurt too much. What was I netting? I did not know the species, did not want to look them in the eye.

SPECKLED DACE

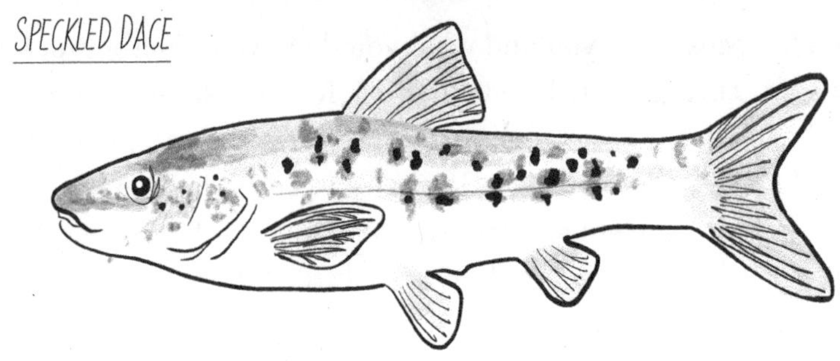

SPECIES	*Rhinichthys osculus*
NUMBER OF INDIVIDUALS, SITES 1–9	0
DESCRIPTION	Their mouths are pudgy lipsticked frowns. Their snouts and cheeks are freckled. In the short video I watch after our survey expedition, they are golden beige and ghostly, more like light than like substance.
FIELD NOTES	We find none.

What happens when you don't find a thing at all in a survey? The Gila topminnow—the original fish that had sent me looking for Peter Reinthal in his office—has long since disappeared from Aravaipa Creek. Did our results mean that speckled dace are gone, too, irretrievably?

Earlier, on our way to Aravaipa in the white university van, we rattled down unpaved roads toward the creek. Dust feathered out over roadside flowers and the biologist driving the van—a mammalogy professor helping out with the expedition—quizzed the students as she glared over the wheel at the rutted route.

How many species of native fish still live in Aravaipa Canyon? she shouted over the rattle.

Seven! responded the undergrads in the back seat.

If we don't find something in a survey, does it mean it's not there?

No, sang the chorus.

That winter, I flew away without answering. Outside Fairbanks, I curled up in a friend's cabin and let the silvery winter wash my lungs raw. Some mornings, my friend would leave in the dark to train dogs, and her girlfriend and I stayed back in the warm cabin: cooked noodles, shared favorite poems, stoked the fire. Other night-black mornings, I would rise to help my friend scoop frozen dog shit out of the kennel yard. I learned how to harness the dogs: *Let their elbows bend the way they want to—don't force it*, she demonstrated. Later, we hitched up the dogs and flew through the leafless woods in the low slant of sun. She showed me how to steer the sled, how to set an anchor to keep the dogs from running too soon.

Out there, on the white trails, she told me, *Some mornings she and I wake up loving each other. But some days I'm only attracted to her as a friend. We just hope that we don't both stop loving the other at the same time.* I considered it: the way affections surface, shimmering. The way they slip under the flow sometimes, aren't always findable.

In my dreams, I see them: my friend and her partner wading out at the edge of the river, the places that haven't yet frozen over. Mist rises from the crust of ice across the middle of the river, catching light like thrown glitter. The two women kneel, dip their hands into the water until their wrists are stinging and red. Her partner pulls up fish after thrashing, lacquered fish. My friend comes up with nothing. Where are they? Her partner smiles and wades on. They'll be out here again tomorrow morning, their breath catching in the cold, looking for quick things stirred up from the streambed. Her partner knows that tomorrow could be different. That they'll hold their open palms out to each other, matching fish twisting across their lifelines, and this will be enough: the knowing, the not-always-catching. The current moves past the sharp edges of ice, shifty and alive.

LONGFIN DACE

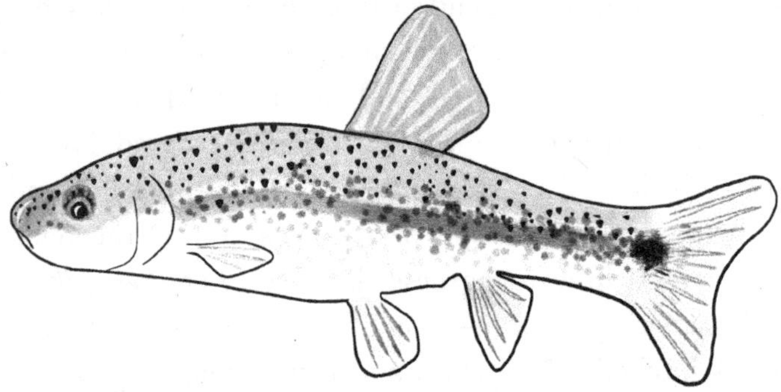

SPECIES *Agosia chrysogaster*

NUMBER OF INDIVIDUALS, SITES 1–9 898

When the longfin dace's stream runs dry, they can survive longer than other fish in the damp algae mats that remain. In this warm chlorophyllic ooze, where oxygen runs low and temperatures run high, they hold still and hold tight till the conditions become friendly again. They are like pearly gems embedded in green fabric.

DESCRIPTION

In the hand, the longfin dace look very much like spikedace: moon-pale bellies, and round eyes set in pearlescent sockets, and thin tail fins like windows through which shows your wet and gritty palm. But where the spikedace has the one pointed fin ray on its back—that identifying jagged fan stave—the longfin dace's fin rays are softer, bendable.

I grasp them with my thumb across their backs the way that I might hold a lighter. The difference between the dace species is subtle, subjective. Is the fin ray sharp? Is it not? I hold one and try to pay close attention. The fish is like cool oil in my hand, and the fin rays are stiff and slimy but maybe not sharp.

FIELD NOTES

One longfin? I say with an upward lilt of uncertainty.

Peter, at the survey clipboard, looks sharply in my direction from where he stands onshore. He isn't going to identify it for me. Well? he says impatiently. Name it, he says. Commit.

One longfin dace, I call back. No one checks my work. I slide the fish into the bucket and reach into the net for the next.

I returned to Tucson that winter. Though I avoided the conversation, the rollerblader sat me down and asked for an answer. Were we in a relationship? Did we wish to continue to be?

I thought of our survey buckets at Aravaipa. How, when the fish had been identified and counted, we spilled them back into the stream. How the empty buckets clattered against our thighs as we walked. As we hiked onward, the fish we'd freed went flashing across the creek bed, gone back to their lives so different from ours.

I curled up against the rollerblader. I knew that our bucket—so to speak—wasn't empty. Though I didn't know the first thing about categorizing whatever slippery things lived inside it, I wasn't ready to spill it dry.

I tried to explain to him why commitment was tricky. I fumbled, telling him for the first time about the broken engagement, about the things that hurt. Hadn't we already been going through the motions, without needing to lock ourselves into a neat and nameable partnership?

This I knew, though: I wanted to keep growing, keep kissing. I was willing, I agreed, to categorize ourselves for the sake of continuing to move with him, side by side. I was ready to try.

I care about you, I told him, though I never used the word I think he hoped for. Neither did he.

But it was enough for him, for us, for then. A conversation, an affirmation. And so we waded on together into the springtime, though unnamed things still rolled and glittered in the shallows ahead.

As the survey team waded downstream, we came across some hikers moving the opposite direction, sloshing in from the west.

They eyed our buckets and our nets and asked what we were up to. We described the survey, and they nodded and apologetically said, *You might be out of luck, we haven't seen a single fish yet today.*

Things that are fishable are not always seeable, I wanted to tell them. And sure enough, at our next study site, where the hikers' footfall had only minutes before set bottom silt and mica swirling into the current, we pulled up several dozen invisible things. They struggled and sparkled in our buckets and, when released, became shadow again, became streambed. And in the downpush of water and the stomp of heavy feet, several dozen other invisible things—hovering and pebble-smooth and wing-finned and wild—evaded our nets entirely, no doubt.

I think about the hikers, still. I think about their journey upstream, unburdened by the knowledge that fish lived in the creek at all. I think about how they must have looked up at the pink-orange cliffs and listened to the watery, descending call of the canyon wrens as the afternoon deepened into dusk, and I think how, all the while, a thousand fish swirled close enough to touch, knocking quietly against shin and sole.

I think, also, about you and me. I think about the places where it startles up to the surface sometimes—that slipperiest fish of all. Out of the green-gold and quiet current that we share, I pull up my net and there it is. I find it under the latticed tree shadows as we go biking along the north edge of campus. And again, in the silky cotton of your blue bedsheets where your hand floats in sleep toward my belly, I find it—distinct and muscular and unmistakable. We go walking after dark up a canyon outside of town. We slip loose of our clothing and splash, yelping, into the cold winter water, and afterward, on the polished rock next to the pool, we lay shivering under the stars, skin on skin, cool and slick as scales. And there, next to us, flashing in the proverbial net, I draw it up again. I feel the shape of it, the slight curve of its body against the curve of my heart, the small and important ways that it is different from all the other species I have netted and set loose.

Name it! Peter had shouted. Someday—soon now, I think—I will.

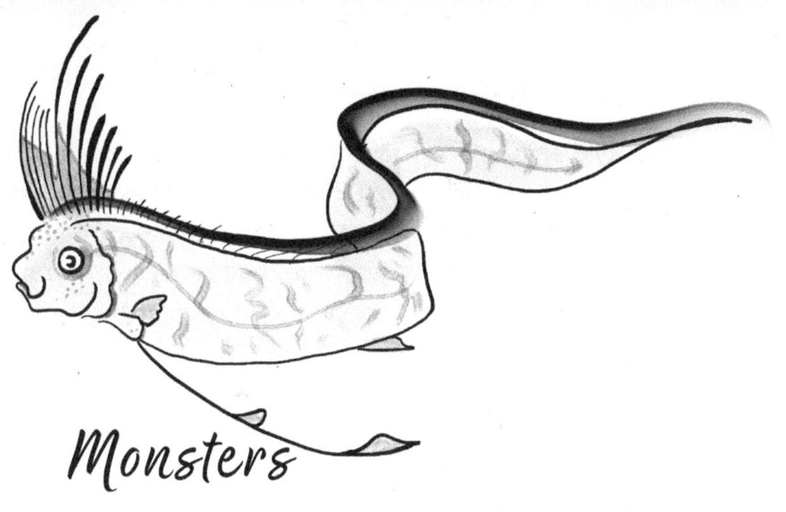

Monsters

WHEN I FIRST MET a sea monster, I was wading in the shallows on a desert island off the coast of Baja California, pushing a kayak into the water for a pair of passengers who had come ashore from a ship I was working on as a naturalist. The bay was curved and calm, but where the water met my calves, it began to roil, and just below the surface, I saw two of them: serpentlike bodies, as wide as mine, barreling straight for us like Nemo's submarine charging oceangoing vessels in the opening sequence of Disney's *20,000 Leagues Under the Sea*. The monsters clattered against the side of the kayak and thrashed in the turquoise shallows. I knew exactly what they were. I abandoned the kayak, running after the fish like a girl chasing a unicorn.

Oarfish! The stuff of legend. Open-ocean animals that rose to the surface almost never, with backbones as long as buses, with flickering red plumes arcing from their heads, with bodies like snakes. These were the inspiration, some think, for sea serpent sightings aboard long-ago schooners and brigantines plowing toward unexplored continents. I had, tucked on a hard drive somewhere, a picture of twenty-five US Navy SEALS holding a giant oarfish aloft after they'd found it beached near San Diego in 1996. I've always loved the possibility of monsters in this world—things that people haven't yet labeled and filed into textbooks. The oarfish was as close to a monster as a person might hope to get. The first live footage of one had been captured only a couple of years earlier, in the nearly lightless twilight zone of the Gulf of Mexico. Most of what we know about oarfish we've learned through accidental catches or occasional dead bodies washing ashore. And suddenly, in the middle of a workday, I found myself chasing after two live ones through the shallows, calling out to my coworker, salt spraying up across our calves as we raced to see the monsters more clearly.

The oarfish were in a bad way. Something had shaken them from the deep water where they lived. When we reached them, they were nosing up against the sand, muscling themselves out of the water. My coworker and I knelt in the surf and pushed back. We gripped the bodies of the fish and heaved. We waded with them out toward deeper water, salt sogging our shorts, spray frizzing our hair. The fish resisted, turning persistently toward shore. Their mouths, little silver frowns, opened and closed. They drove themselves onto the beach again. And all we could do was sit with them, our bodies so small next to their great glittering lengths. We watched their colors fade from flashing blue and pink to gray.

I understood that this was an astonishing encounter. So much of the lives of oarfish—their diets, their movements across oceans—was still entirely unknown to science. When they finally grew still on that long crescent of beach, I radioed the ship. *Can we bring one back? Can we count its stomach contents, learn where it's been?* But the captain was concerned that we might break some local fishing law by bringing them aboard. So we left them there, going dull under the sun, and we gathered our guests, and we departed for the day's hike, leaving the beach behind.

Unable to do my own dissections and still dizzied by the hunger to know things, I sought out an ichthyologist later that year at an academy in California. I arranged a meeting with him, the world's foremost expert on oarfish. I wanted to hear his stories, to unpack the mystery of these monsters. *What good timing, we've just been given a specimen*, and he pulled a smallish oarfish from a white bucket filled with pungent preserving fluid. The fish was brown and soggy and smelled like a bad martini. The expert had been at work, taking small slices of the fish—scales, tongue, teeth—to examine by microscope. Much of what he was finding was unprecedented. The oarfish that he was studying was different from all the other species of fish that drifted in their ethanol baths on shelves and floor corners in his workspace. *I don't see any squamous epithelium!* he exclaimed, showing me a slice of mouth under the lens. *This is beyond my experience.* He chuckled at the absolute newness of it all, and I marveled at being witness to it, live and in person.

But he had odd enthusiasms, too—ones that made my skin prickle. At the time of my visit, Donald Trump was the new president-elect, and as our conversation deepened, the ichthyologist looked up at me from across the jars of dead fish on his office table. *I don't know what all the fuss is about this cunt march*, he confided in me. Women were marching in the streets that week, rallying against a new leader who might soon strip them of the right to choose what could happen to their bodies. Pumping blood heated my ears, but I steered the conversation back to fish, jotting down notes, straining to be a dispassionate journalist. Later, unbidden: *I hate all Muslims and if you know what's best, you should, too.* My belly felt hollow. He tried to talk me through his list of reasons, and I guided our interview back to migrations, matriarchal lineages, diet. When I suggested the possibility of climate change as a driver behind increased oarfish strandings, he scoffed, *There's no evidence for global warming*. Was this the purview of a specialist, to focus so raptly on a solitary thing that the calamities and interconnections of the bigger world drop away, unseen? I snuck a glance up at the larger laboratory—office spaces with fish posters on walls, researchers shuffling from the copy room with loose-leaf printouts in hand. Did the ichthyologist speak this way to them as well—with dogmatic bluster, with irreverence for fragile things? Still, I pressed my molars against one another and followed him through his workspaces. I should have been brave, should have spoken my mind. But I was scared, and young. And I refused to leave until I had learned everything I could about oarfish. To win that privilege, I needed to stay steady, neutral, placid as the currentless abyss.

We moved together into the room that housed the scanning electron microscope. He sent the lab tech to fetch the specimens he planned to process today. She, a young Latina scientist, came into the room with a handful of small boxes and, like me, she kept her mouth closed as we laid out the boxes. But when we made eye contact, her gaze felt grim. Maybe she, like I, had been taught to see science as a quiet meeting space, a laboratory in the mind where ideas gathered, where careful method guided the search for true things. But something else had walked through the lab door, something mean and human and a little monstrous. What is a monster, after all, other than an unfamiliar thing moving through a place we thought we knew?

Here in the lab, tiny specimens—fin rays, oarfish scales, miraculous little cave fish the size of my thumbnail—had been sputter-coated in gold. We would run narrow beams of electrons over them, which would ping off atoms at different levels in the samples and yield three-dimensional pictures of the things our eyes could not see: images detailed enough to see a single hair's fray or the honeycombed dimpling on a grain of pollen. One by one, we sent the tiny fish fragments under the beam, each bathed in precious metal, locked up tight in a golden straitjacket, for science, far from the living sea.

I was glad, when finally I left that lab space in San Francisco's growing dusk. I was glad, remembering the beach, the thrashing water around my calves, the two monsters laid out like fallen gods on the sand. Glad I had not cut them open, to count and quantify. I was glad I'd left them whole, as I moved away with mystery still alive in my chest, as I looked back over my shoulder to watch the gulls sidle close and pop the lidless eyeballs loose from their sockets, as I walked inland and left the desert to work its quiet, dissolving magic on those bodies too sacred for science.

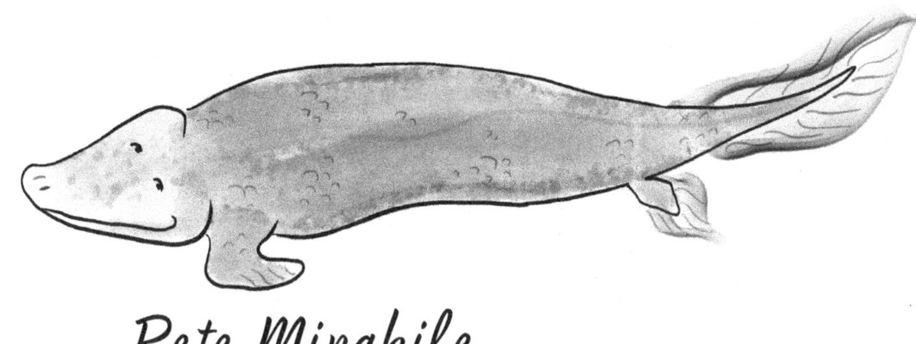

Rete Mirabile

I am out with lanterns, looking for myself.

—Emily Dickinson

INDOORS ON THE PADDED table, I grip one hand with the other, squeezing my thumb, trying not to fidget or to yelp aloud. I'm stretched out on my belly across one of those long strips of medical parchment—the kind that sticks to skin, the kind that punches through in the thin places where your body presses. *Let me know if you need a break*, the laser technician says, but *I'm tough*, tell her, and I grip my thumb tighter. The air smells like a poorly wired hair dryer: warm cloud of scorch, the sizzle of burning carbon. My legs sting where the technician works her rod across my shaved skin. Red bumps prickle open like starbursts. She tells me that's where the laser is most effective—the reactive spots, the places where heat scorches the follicle all the way down to the dark and jellylike root.

I hold tight. Vanity dragged me here in the first place, and now it keeps me from squirming. I imagine my hairless legs.

Clear water streaming across bare body as I take my weekly shower on the hill above the glacier where I spend my summers—not needing to reach for a razor, not loitering among the mosquitos any longer than necessary. Smooth skin on smooth ankles as I lace my boots and duck under the zip entrance of my wall tent. The ease of nakedness. And the purifying burn of it, too. As if the hair is an inessential layer to be shed, in search of something true and whole beneath it. It is an unmaking of sorts—scorching away the mammalian parts of myself, imagining what vital core might remain. On the table, the laser whines and a stream of cool air whirs across the singe.

Before I ever thought much about what makes my body a warm and dark-furred mammal's body, I learned what makes a fish a fish. In a fabric-bound turquoise notebook, I sketched bodies with fine-tipped pens, watching my professor draw them on a board at the front of the room and doing my best to copy what I saw: snaggly sets of pharyngeal jaws deep inside throats, longitudinal blocks of red muscle, facial canals laced with electromagnetic receptors. It was my first introduction to taxonomy—the discipline of defining and dividing groups of living things based on physical features. Odd and wondrous bodily diversity bloomed across my pages. My own flat molars and dull senses seemed boring in comparison.

But I also began to apprehend our similarities, bone-deep. Kingdom: Animalia; phylum: Chordata. In my notebook, I sketched structures that could have been human if I weren't carving them into fish-shaped spaces. Cranial cavities, spines threaded through with nerves, eyes to see, hearts to pump. In that classroom, I saw that the coat-tree of life is draped heavily with shared and overlapping fabrics, traits that fold distant relatives together.

Getting naked didn't always burn. It used to be easy. Summers in my late teens, living at a national seashore, I'd join the other counselors in the woods on weekend nights after campers had left. We'd peel away our clothes and go wandering in the salt fog, under the buckeyes. Nettle would brush our ankles and mud would cool the sting. Young women, all of us, we were still learning our curves. We walked one after another along thin trails. Pennyroyal and ceanothus blooms spiced the night air. Grasses slipped like silk against dark hips. Starlight caught our clavicles, the knot of fur where our legs met. Our nudity was uncomplicated and sensuous.

At what point, in the years that came after, did nakedness get difficult? I remember letting a new man look at my body after I left my broken engagement. The kissing was easy, eyes closed, brick wall rough against nape, cigarette smoke pluming through New York's winter air. The seeing, later that night, was harder. I feared that my body, if seen too clearly, could never be loved again: its small constellations of moles, its green-veined mapwork. In the years that followed, I left and left and left, before anyone could perceive me too deeply and judge my worth.

It's a chilling nakedness to be seen, to open like a book and hope that someone likes the words within. And so I collected my little losses in private. Unfaithful fiancé, unloving kisses from uncaring lovers, unbreathing dad gone lifeless in the cool drift of kelp a hundred feet below the surface. At what point did I lose the shape of myself, beneath my baggy histories?

Taxonomy is the science of naming, and also the science of separating. All that semester in my ichthyology class, I circled the systematics of identity. What makes a fish a fish and not a human? Beyond muscle and rib, what differences make an animal one thing and not another? In order to arrive at the essence of "fish," we needed to winnow away the shared things. Fish have scales, sure, but so do reptiles. Fish have a beating heart, yeah, but so do we. Taxonomically, a thing is identified by its unshared characteristics—the traits that it possesses that nothing else does, or at least in combinations that other animals don't share. Like us, a fish has a brain housed in a protective skull. But unlike us, cool blood circulates through its muscles, matching the temperature of the water around it. Instead of legs, a fish's limbs are shaped like fins, either embedded with dense bone or fanned through with delicate rays. These aren't attached to the rest of its skeleton but instead they connect by muscle to the body of the animal, each a free-floating bony inflorescence like a corsage clipped to the outside of a garment. A fish gulps water through its mouth and funnels it through its pharynx and out through a series of fringy gills. Blood passing through tiny capillaries in the gills absorbs oxygen from the current and flushes the fish's body with red and pulsing life. Just as a mammal is known by its warmth, its milk, its hair, its lungs, a fish is known by its cool blood, its gills, its limbs in the form of flowery fins. Each of these identifiers, I drew in my notebook. It was tidy and pleasant and mathematical, to lay life out like this: black gill filaments in a flat white sea.

My mother and I strip away our towels and wade, steaming, into the hot springs. Mist lifts and settles on us in warm beads. We seat ourselves on the stone benches, arms floating and fingers loose at the surface. Her body is thinner than mine, and silverer where the water lifts and drifts our hair. I can see my body in hers, though. Torso long like a tree, veins stitched across belly like a diaphanous green net. Our bodies bend, too, under the same weights. One loved one drowned, another trapped inside an addiction that leaves us wondering each morning if today will be the day that his neighbors find his body, alone and cool to the touch, in the duplex where he lives. Father, husband, brother, son. Our robes are woven out of the same heavy material.

Though my skin touches hers sometimes when the water rocks us, I know that we aren't fully naked in front of each other. What would it take to lift the leaden layers? Sometimes we shout at each other. Sometimes we call each other names, and sometimes we cry, and sometimes we read stories aloud to each other like girls in a blanket fort. *I'm sorry*, I tell her today, and *I love you*.

Beyond us, in the river below the hot springs, fat catfish gather in the warm outflow. Their whiskers waft backward in the current; their bodies slide against one another.

Taxonomically, fish are tricky. There are basic things that make a fish fishy, but those can be peeled away. One by one, the identifiers that I illustrated so cleanly in my notebook proved fallible. Not all fish live in water. Mudskippers with their sky-blue freckles and their froggy eyes and their long, elbowed fins only duck underwater occasionally to moisten their gills. They spend the rest of their lives on land. They can drown if submerged for too long. They waddle across mud flats like pudgy little salamanders and climb trees and don't worry too much about taxonomic propriety.

Not all fish have scales. Some are gelatinous and smooth. Tiny tide pool sculpins—with bodies the color of barnacle, of mud, of mottled stone—move through the world with no armor. Their skin is slippery and naked against the edges of the world.

Some fish breathe air. Since the Triassic, lungfish have slid through swampy shallows and small creeks, rising to the surface to pull air into their lungs. If their home dries out, they'll burrow into the mud, leaving a hole for air to enter. Tail over head, they'll wait out the drought, sometimes for years.

Some fish have warm blood instead of cold. The opah, with golden eyes and a body flat and round as the moon, flaps its fins to keep warm. It is endothermic—heated from within—and though it swims in deep, cold currents, its body stays warmer than the dark water around it.

So many essential things can be stripped away and a fish still remains, more or less, a fish. Despite the exceptions and anomalies, I suppose at the end of that semester we must have arrived at some overarching definition for "fish." I do not recall what it was.

I have tried, of late, to get naked more often, and more bravely. I burn away my hair. I dance on go-go boxes, where older women tuck wet bills into my boots. I let reality TV producers fly me to LA, where they try to convince me to spend twenty days in the wilderness without clothes. (I consider it; I say no.)

I'm not sure what I'm stripping toward. I don't know what, or whom, I'll find beneath all the layers. I'm as untidy and mismatched as a sock drawer. Sometimes I plant heirloom seeds and sometimes I eat Cheez-Its for breakfast. I go walking in the woods in search of mushrooms and wild rosehips, I walk on stilts at music festivals, I shop for extravagantly sequined dresses at thrift stores, I hike into the desert alone at night so that I can sleep under a quiet sky.

I sit on café patios, where I slip crumbs to birds and watch them flit and fluff themselves. Sometimes I wonder if I like them more than I like people. I've hurt and been hurt. I believe in love.

Sometimes I go for weeks without thinking about my father. Sometimes—when I see a train rainbowed with graffiti, or a chipmunk that he surely would have tried to feed—the memory of him descends like sudden rain and, even years later, it's all I can do not to lie down flat in the cloudburst of hurt, right there out in the open.

I'm not sure what it all adds up to. I peel away my multitudes like coats. If I shed some vital layer, who might I lose? Who might I become?

I soften my gaze. I imagine layers of muscle and bone drawn on a clean white page. Where, between jaw and spine, hair and finger pad, are the lines that make me identifiable? The taxonomy of identity is as refractive and hypnotic as a fish in motion.

In the evolutionary story of fish, a transitional fossil is one that has defining traits of both an ancestral group and the group that follows it on the tree of life. Consider fish and the four-legged amphibians that descended from them nearly four hundred million years ago. There wasn't a clean divide: a fully formed tetrapod hatching out of a fish's egg. Instead, evolution zigzagged and inched toward newness. There were in-between creatures that bridged the gap. *Tiktaalik*, Neil Shubin's arctic fishapod, is one of those transitional fossils. It is a fish, kind of, with scales and gills—but, like a tetrapod, it has a flattened crocodilian head and sturdy bones for walking rather than swimming. Strip down its characteristics, one after the other, and still at the center of that long-ago animal there's no tidy identifying feature that would allow paleontologists to place it in one taxonomic clade or another.

The fact of the matter is, fish don't fall into their own separate taxonomic clade at all. We thought they did. We called them *Pisces*. And then the fossils began to reveal a trickier story. If the tree of life really were like a tree, then a clade would be like a flower, a complex one that blooms open into a thousand little starburst buds. The clade, like a flower, shares a common stalk—a common ancestor—and the compound blooms that divide up out of it are the descendants of that ancestor. Each clade a separate stalk with its own many-branched umbels. Fish all share a common ancestor. But from lobe-finned fish descended reptiles, mammals, birds, amphibians. Because of this, fish can't be categorized under their own distinct taxonomic class—we're all part of the same evolutionary flower. And the lines between us are not crisp. Transitional fossils help remind us of this: of the in-between places. They help remind us that the divisions we've imposed are arbitrary, that we all belong somewhere on a family chart that looks less like a chart at all and more like a continuum of variation. We are not an "either" or an "or." At our most naked, we sizzle with connectivity.

In a music venue on the west edge of downtown Tucson, I sometimes take my clothes off. It goes like this:

The stage stays empty as the low electric guitar opening of Gin Wigmore's "Kill of the Night" thumps from the speakers, loud and scratchy. I give it a few measures, letting the anticipation ripen, and then I sidle from the wings, head tucked down, leading with my glittery boots. I am dressed like a big-bosomed shark. I wear a fleece onesie—a pajama suit, essentially—with a toothy hood and gray fins. I stroke my hind fin, mosey to the front of the stage, and make a show of peeling off one fleece sleeve and then the other, whirling each before tossing them behind myself onstage.

The danger is, I'm dangerous, and I might just tear you apart, Wigmore croons, and the guitar breaks into a quick pulsing riff. I tug at my shark zipper, inching it open from the neck in a manner that I hope appears coy and provocative. Maybe the audience hopes for cleavage—dollars fly—but instead, I pull an enormous stuffed goldfish from its hiding place between my tits. The front row goes wild as I swing the fish high for all to see, lick it suggestively from tail to snout, then give it an abrupt shake with my mouth and discard it.

It's a new discovery—the playful nakedness of burlesque performance, this glittering, giddy irreverence. As my sequined layers fall away, I feel like I'm accessing some younger version of myself: a girl who loved theater, a girl who wrote stories, a girl who learned clowning and costuming at summer circus camp. Joy sizzles up, right near the center of myself. This nakedness is for me. But a boyfriend looks at the pictures from my latest show. *Can't you be classy?* he asks, and the joy gutters. *This is vulgar. I want the outdoorsy Hannah back.*

But maybe I am *Tiktaalik*. Maybe I'm neither what the boyfriend thinks I've become, nor who I used to be. Maybe I'm both. The boyfriend leaves me. I stay onstage, hunting that elusive sizzle, stripping closer to some unseen heart.

Strip away every layer of a fish and inside you'll find, scooped below the spine, a balloon-like organ. Shaped like two sacs, it fills with gas through a spongey gland nestled above the vital organs. Woven from tightly twined fine veins, a rete mirabile—a wondrous net—feeds oxygen into the swim bladder and absorbs it back out, according to how buoyant the fish wishes to be. The swim bladder lets a fish balance in the water, keeps it from sinking too deep or rising too high. It's a stabilizing organ.

But a swim bladder can also make a fish vulnerable. If a fish rises too quickly in the water—pulled up in a net or on a line, dragged up out of a lightless deep current and into some other world altogether—the change in pressure might kill it outright.

Expanded bladder, popped like a too-full party balloon. Pink tissue, taut and sticky, pressing against spine and kidney, pushing up out of the throat like a birth gone wrong.

Mostly, though, swim bladders are evolutionary wonders. Without them, fish likely wouldn't be so diverse, making up more than half of all known vertebrate species. Swim bladders allowed fish to take new territory—allowed them to regulate. Fish diversified into far-flung habitats because of their ability to sink and to rise.

Because of swim bladders, fish are multitudinous, cosmopolitan, complex.

Because of swim bladders, fish are buoyant.

What small and irreducible core will remain of me when everything else is finally stripped away? A raw little self—volatile, vulgar? My naked teenage self in the woods, toes in the cool mud, feeling for the trail? A bright orb of pain, throbbing with loss?

Together, a friend and I fill my bathtub with warm water and, as it fills, we toss flowers across the surface. Long lilies. Fat sunflowers. Baby's breath, diffuse and papery. The water pools in the black centers of the sunflowers and carries torn petals eddying along the edges of the tub.

We step into the rising water. We lean back together: heel against ceramic, thigh against thigh. Wet blooms knock and drift against our legs. The air is filled with the spice of eucalyptus and the soggy sweetness of warm flowers. Our clothes cling, and so we pull them away. Our tops peel back like skin. We rise and step out of our underwear.

As we sink back into the bath, ragged flowers roll across our bellies. Petals hug our breasts. We sit up and lean toward each other. She pulls me closer. Her breath is humid, and her lips against mine are as wet and naked as carnations.

Her body and mine are warm and downy, curve sliding against mammal curve. We kiss, and our fingerprints press into each other's naked skin, little stamps on a letter still being written. She grips my wet hair; I let my nakedness soften me. I've never loved another woman—someone so much like me: brown fur, hips soft as animals. Looking at her body is like looking at my own, and my gaze gentles. There's weightlessness to it all. If hurt and guilt and ambivalence are flowers, they are absent from this current we've drawn around ourselves today.

It's just us, plucked clean from the tangled branch that grew us. It's just tenderness, just lift. How silly to think the measurable body held answers. Beneath my twenty-four ribs, beneath a warm sternum, is something else entirely, undiagrammable.

It tugs toward joy.

All along under the layers was some ballooning buoyant organ tucked there right at the center, against the spine. Together, among the shed petals, we float.

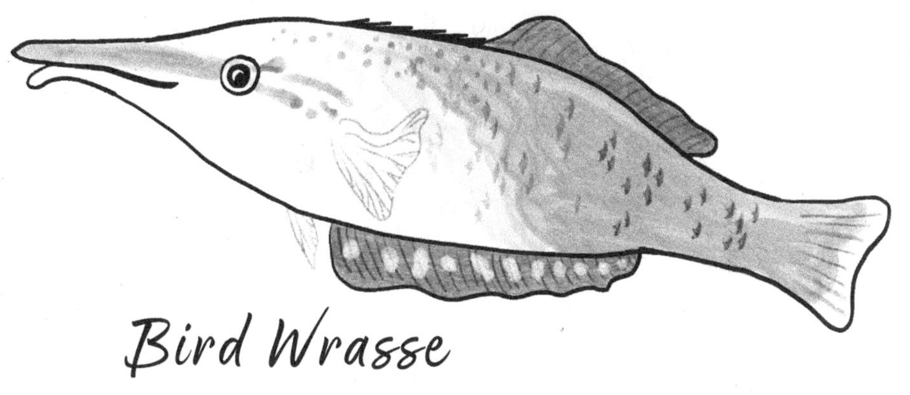

Bird Wrasse

BIRD WRASSES, *Gomphosus varius*, are perfectly common. They wander the waters of the Indo-Pacific, from the warm eastern edge of the Indian Ocean to the flung crumbs of Hawai'i's archipelago. They spend their days solitary, snooping among coral for small meaty things. Their snouts are long, like bird beaks, which allows them access into tight crevices and inviting nooks where their food nestles. By night, they tuck themselves into snug spots in rock or coral and they snooze. The males are a deep and brilliant green, like emeralds, with a bright flash of lime at the shoulder and faces as blue as if they'd just plunged headfirst into blueberry pie. The females are drab. They sell for half the price of their male counterparts on online aquarium stores peddling wild fish. Their bellies are pearly white and speckled with tiny black spots that darken at mid-body so that their rear halves are nearly all black—but, unlike the males, their beaky snouts are bright orange, making the female the birdier of the two sexes, indisputably.

My fellow snorkeling guide rolls his eyes at me when I go following a bird wrasse low among the corals. I hold my breath and trail her like a lost soul follows swamp lights. He shakes his head when I spring to the surface, exclaiming my discovery. He is looking for ribbon-soft eels, for quiet sharks under ledges, for the *lauwiliwili nukunuku ʻoiʻoi ʻeleʻele*, the long-nosed butterfly fish that shows up, all black instead of its usual yellow, in just a few places along the Kona coast. He likes the rare things, the fierce ones.

But I am undeterred by his indifference. The bird wrasses are, to me, numinous. If kept in lidless tanks, they'll fly right out toward freedom with swift birdlike flaps of their fins. Nobody knows how long they can live. They are unholdable, unmeasurable.

And, when no males are present, a female might undergo an extraordinary transformation. Enzymes sparkle awake inside her body. Her ovaries degenerate and crypts form there, where eggs used to be. She produces sperm, which moves through her oviducts, travels down passageways that are not wholly male but no longer female, either. She loses her orange beak, her speckled skin, and—changed, brilliant, unbeholden to her past, unassumingly miraculous—she wings greenly away into some unimagined new life.

Watershed

IN THE MIRROR, MY veins show through my skin like green rivers. They glow. They fork up my underarms, across my breasts, and I am like an overhead transparency, some other nation's watershed projected across myself. I try to feel them—my veins—where they come from, where they're pulsing toward, but the map is foreign, unfeelable, tucked just beneath my surface.

A watershed designates an area that shares a common drainage—all the high points sloughing snowslush and dust and rain down toward a river basin. A river might begin in cloud forest, where minnows dart under vines and bullet ants march along the banks in their shining black armor. It might not be more than a runnel at first, a rivulet, moving like thin glass over fallen leaves, under humid banks brown with loam. It might meet other rivulets, might join forces, might bristle with white rapids. As it collects warm rainfall, a river might turn strong and brown and knotty with uprooted jungle trees. It might rise above the banks, might swallow an oil dock, might swallow a child. Where river meets river, freshwater dolphins might spin in the quick places where different-colored currents knock into one another. Dark fish, long as crocodiles, might hover in the eddies.

Or a river might begin as snowfall, might race seaward past flocks of ice-white geese and over the backs of sturgeon old as dinosaurs, might tumble into a gray bay.

A river might tire of its old trail to the sea, might change things up. A river might sidle to one side or the other and settle into a new bed, leaving boats dry on their moorings, rocking atop the crackled mud. River towns might wake up to find themselves riverless. A river might change its course a thousand times, leaving behind bed layered atop old bed like snaggled yarn dropped across a table. A geologist's depiction of the Mississippi River looks like this—long-dry channels drawn in red, yellow, knitter's blue. The mapmaker bored into the ground sixteen thousand times to find where the old beds lay. He illustrated them one atop the other atop the other, those old tracks of the wandering water, looping and purling around the heart of the still-living river, which, at the center of it all, still flowed north to south to the Gulf of Mexico.

And then, too, a river might bubble out of a desert spring, might pour like syrup under sycamore boughs and past foraging coatimundis in cool pink canyons, might water cattle, might water pecan orchards, might water cities, might never reach the waiting sea.

Do whom do the rivers belong, these yarny veins, these minnow condominiums, these corridors for migrants of all species? In New Zealand and in California, there are rivers that belong to themselves—or rather, rivers that share the same rights as humans. They can't be dammed or fished or diverted without legal representation, without proof that the river won't be hurt. Not so in the Southwest, where the Colorado River runs dry before it reaches the sea. Where 90 percent of Arizona's waterways have disappeared in the last hundred years. Not likely to return, says the informational placard at the Arizona-Sonora Desert Museum. The display shows two watershed maps of the state: before, after. Whole tapestries of water gone missing.

Can rivers just really walk off and leave? Where do they go? Is it better, there? In the warm and pulpy bellies of cows? In cool turquoise backyard pools? In the underground chambers of shallowing aquifers, echoey and dripping?

In Arizona, about ninety thousand miles of water trickle and crisscross over the state's rocky frame. They roll across my tongue with names that denote pounding storms, deep time, glinting fish, salvation: Thunder, Fossil, Silver, Bright Angel. The veins in my body travel the same distance that the creeks do, splitting and looping and stacking above one another for ninety thousand miles like layers of lace. I think about disappearance. What does it feel like? I think about those thousands of pulsing miles, unfeelable just beneath my skin. I imagine a before and after map of the body: arms carved open, 90 percent of the veins sliced loose like the branching tubes that you sometimes pull from a warm carnitas taco. I imagine them gone, set aside to drip dry. How long can a bloodless body live?

Lazarus in the Desert

BUT LISTEN: THERE ARE rivers that return.

Once I followed a long corridor beneath the ground floor of a University of Arizona science building and cracked open a door where a handmade sign read GIANT WATER BUGS ON DUTY. In this underground lab, amid microscopes and screens, a grad student named Drew showed me an hourly chart of the water flow in Tucson's Santa Cruz River. Like a spiking electrocardiogram, the graph peaked with morning bathroom visits and evening showers and, while the city worked by day and slept by night, water levels fell. Tucson recently began funneling treated wastewater into the Santa Cruz, and now the river rises and dips daily like a tide. There are thirteen such rivers in the state—all effluent-dependent, all recently returned from the dead.

And in the Santa Cruz, a tiny silver fish—missing for seventy-five years—has returned, too. The river ran dry in the middle of the twentieth century, and Gila topminnows ghosted. But when water began to flow in the long-parched channel in 2017, these endangered little fish found their way home, all on their own. Nobody knows how, but, like Lazarus rising, maybe miracles don't need an explanation.

When I go walking along the wet banks of the Santa Cruz River, among bent shopping carts and yellow monkey flowers, among flame skimmer dragonflies and cattails and sun-brittled Coors cans, I can't help but think of resurrection, and the shimmering nature of time.

As with fish, so with love. My dad, who never got to see me graduate from college, could never have imagined my bush plane flights over the Wrangell Mountains, my late nights standing watch as our ship spun slowly at anchor in the soft Alaskan rain, the distant and perfect July day under the redwoods as I stood in front of my new husband with flowers in my hair. But weren't each of those moments a resurrection, of a sort? My dad, alive and one stride ahead of me as I stepped down from the cockpit of the Beaver into the glacial wilderness. Alive in the quiet engine room of the sleeping ship, among the pistons and circuitry that he loved so much. Alive and close by my side as I walked down that aisle of fragrant duff on my wedding day, a golden necklace from him glinting at my throat.

Arizona's resurrected rivers are complicated. In the outflow of urban wastewater, we've brought whole ecosystems back from the dead. But in this dry state, summers are heating up, aquifers are shrinking, and the water of the Colorado River—funneled across Arizona along hundreds of miles of cement diversion canals—draws ever lower. Will we continue to recharge urban rivers when lettuce fields in Yuma need irrigation? Once we have brought a fish back to life, what do we owe it? What do we owe the river in which it swims?

The future, of course, is as slick and unnettable as a fish.

But the past? Water it and the luminous ghosts return, return, return.

A Gray and Rocky Place

WHEN JUAN CUEVAS WAS a young man, he traveled by freighter away from the Land of Eleven Rivers. Behind him, opium poppies nodded in the Sinaloan wind. Tucked in his pocket, possibly, was a little glass damiana bottle in the shape of a woman, filled with a few sips of the herb-based love potion that artisans concocted in his hometown. He had heard of the remarkable fishing opportunities on the faraway Baja peninsula, and so, hopeful, he crossed the Vermilion Sea. Along his journey, windstorms churned the water and sperm whales rose to the surface with pillowy sideways blows. Dolphins flanked the freighter and danced ahead of it. When he arrived in La Paz, Juan Cuevas wandered among long-legged pearl divers, and lovers out strolling the seaside *malecón* in the pink evening, and dusty rancheros with crinkly smiles, descended from the Bedouin nomads who had arrived centuries before with the Jesuits.

But Juan Cuevas was a fisherman, and this desert city was too big and too busy for his heart. He set out in search of a place where he could fish for the shark livers, which drew a good price, where he could live away from crowds, where he could boat out to meet the mother ships that traveled from fishing camp to fishing camp to buy up their catch. He found a boat for himself, and he struck out for sea.

First, he traveled north, to the estero at the southern end of Isla San José. But his camp at the mangrove edge of the sea was plagued by little *jejenes*, which bit through skin and drove him wild with itching. The seawater only worsened the itch; the bites grew like constellations along his feet and legs, bubbling and scabbing over like collapsing galaxies.

And so Juan Cuevas fled in his boat, traveling south. At Isla San Francisco, where the bay curved like a horseshoe and ridges like knives slashed down to the sea, he built a small home for himself on the rocky north shore. He walked to the island's flats to dig pits where seawater could soak up through the low dunes. Here, he shoveled up the layers of salt that formed in the dry desert air, big scoops of glassy crystals, white and pink and gold, and with this he dried his fish and salted it so that it would keep until the mother ship returned. But again, the *jejenes* followed him. Again, they bit deep and relentlessly—little monsters, airborne, unseeable, insatiable. And so Juan Cuevas traveled onward, until, at last, he found a rock rising up from the sea.

It was barely an island—stony and waterless and rough. On the old maps, it was called Isla Coyote, but Juan Cuevas renamed it for himself: El Pardito, which means something small and gray. Here, Juan Cuevas built his home, and the *jejenes*, deterred, perhaps, by the dryness, by the unforgiving rock, did not follow. Here, at last, Juan Cuevas found relief. And so he stayed.

Over the blue-green years, he fished for sharks, and for yellowfin, and for triggerfish in the reefy shallows. He was masterful at reading water, at finding fish. He raised a family on the rock, and they stayed and raised family, too. Over the decades they became renowned throughout the Sea of Cortez as wizards with their small boats, their hand-mended nets, their palms creased with salt and glimmering with scales. Aquariums came to speak with them, to hire them as guides in their search for hard-to-catch specimens. Visiting sailors arrived on their rocky shore, looking for the freshest fish as they explored the sea's jagged islands. When the big ships stopped traveling between islands to buy sharks, the Cuevas family would boat themselves the fifty miles into La Paz to trade fish for water, to glance with big eyes at the bustling *malecón* lined with palms, and then to retreat back through the salt wind toward home.

Once upon a time, flung fragments in a loose-knit nebula began to fold toward one another. Gravity tugged them together into gassy spheres and rocky lumps and molten churning planets: eight of them. The third of these planets was messy and hot, orange with melted rock and reeling from all the chaotic energy of its formation. It spun among a million other loose rocks, and one of these—a huge one, traveling quickly, tumbling away from Jupiter—collided with the newborn Earth and rent a dense chunk of it away.

After a time, the two of them—Earth and the piece of Earth ripped loose—both tidied themselves up, letting gravity tease them spherical and hold them in a loose and orbital embrace. The icy load from that mighty collision melted into Earth's basins. A sea formed, and oozy, wiggly things sparked alive under its surface, near hot vents. Green things exhaled oxygen. Gases circled and clung. An atmosphere formed to hold things together. Earth's moon—cut from the same rock, healed from the same collision—lingered near and tugged at the talkative sea like a child tugging a mother's hem.

In the slosh and sway of this new balance, stick bugs and velociraptors and honeycreepers fluttered alive. *Jejenes* learned to hunger for blood. Mammoths collapsed to their knees, spears swaying from a flank or from the space between ribs. People trembled up onto two legs. They built boats from reeds, watched faraway stars, trudged away from drought in search of home, swatted at jungly bugs, sketched blueprints for glittering cites. And all the while the ground beneath them spun. The atmosphere, like the folds of a skirt, caught rising moisture, wrung it back down across the thirsty hills. Rocks tumbled down from mountains and sifted to sea and rose again as rock. And so, amid all the *gallop* and *go* of time, the planet stayed mostly the same: iron, carbon, salt. A slow view—a wide one, from above—might reveal stillness behind all the bustle, a sort of recirculation, churning slow across the planet like a river eddying. Under the moon's blue light, our spinning rock gave and took, circled and stayed.

Working in Baja, I sometimes passed by the island where Juan Cuevas's family still abides. In those days, I was always between one place and the next—finishing up a winter guiding season, looking ahead to returning to Alaska in the summer; leaving one relationship, one foot already stepping into some new love like the tarot fool, falling. Escape was my forte. When things got hard, I moved. It's easy to imagine why I kept returning to boat work: steady movement churning me forward, even at night the buzz of engines rattling the walls of my bunk while I slept.

So I suppose the Cuevas family seemed somewhat marvelous to me, looking in at them from the outside: some fairy-tale fortress at the edge of the sea where no one traveled much farther than a panga's fuel could carry them. Sometimes, they boated out to meet us when our ship anchored nearby—they'd bring empty jugs, and we'd fill them with fresh water from our ship. In exchange, they gave us yellowtail, which we steamed up for dinner.

Sometimes, I came ashore in a little boat to visit with them. Beto, the patriarch, walked down toward shore, straw hat checkering his face with shadow. He'd twinkle at my efforts at Spanish. We'd walk past the blue-and-white pangas onshore, up past the salting shed, and under a palm-thatched roof. He'd lay a fresh-caught triggerfish—flat like a pancake, stippled gray and black, perfect for ceviche—across an outdoor table and he'd slice cleanly along its fin line, top and bottom, hooking his knife under the flap to loosen the fillet from the skeleton. He'd skin the slabs and chop them into cubes. He'd hand me the pink, glistening meat tied up in a plastic bag. All the while, the gulls waddled along the low cement walls, screaming and laughing and shuffling their wings. They nested near the family's front door and wandered fearlessly among the fishermen.

They've been here longer than we have, Beto would say with a softness I don't often associate with fishermen. *We can afford to share*. And he'd throw them the bits of fish that he wouldn't eat, and they'd clamor for it, a swirl of white feather. He has lived on this waterless island long enough to know about generosity.

Circling above this giveaway, a gull would barely be able to discern the gray outbuildings from the gray of the island. The homes are painted with loving depictions of whales and sharks and orcas, now faded—the soft colors lifting away under sun and wind. From above, a gull might just barely see me waving good-bye, and cradling my bag of *cochito*, and running back toward shore, all velocity and verve. A gull might watch me cut by boat across the turquoise shallows and into the deeper blue, onward and away.

A gull might see Beto turn and stroll back across his improbable gray rock—all iron and carbon and salt. It might see Beto sit for a while in a sunny place, then rise to check on the salt shed, scuffing in his sandals across the concrete foundation, which is all that remains from the first house that Juan Cuevas built. The rest has returned to rock. A gull, from above, looking down on this blue world, this little rock, might see Beto go about the quiet business of staying.

Shark Shadow, and Some Balloons

DAD WAS BELOW GROUND, under the shark tanks when the rattling began. It was 5:04 p.m. on an October evening in 1989, and the sharks and other fish circled the huge roundabout, their refractive bodies casting shadows across my father, who stood alone, freshly arrived for his evening engineering shift in the bowels of the Steinhart Aquarium. Maybe it was the faint buzz of shaking glass, or something crackly shifting underfoot, or a sudden muscular agitation among the toothy swimmers overhead, that first set his skin prickling. The sound was untraceable at first, a soft grainy growl like the crunch of tires across gravel.

Fifty miles north, my mother felt the room moving. For a moment, she thought it might be one of the boys who had come over for play group, thumping the floor as boys tend to be good at doing, but then her bones recognized the tremor as something deeper, jouncing up through layers of mantle and crust. *Everyone outside, now*, she ordered, never one to fuck around when it came to danger.

The shaking subsided, and the children went home, and an uneasy quiet settled over our little town. By and by the neighbor came over the fence, leaned over, and announced, *The Bay Bridge has collapsed.* The enormity of it dawned, but by then landlines were useless—overstrained and thick with crossed connections. My mother rang and rang.

Under the tanks, my dad listened to the rattling grow to a roar. For a moment or two, he wondered if it might be the caterers above, rolling carts across the floor, preparing for an evening event. But his bones, like my mother's, knew the deeper truth of it, and he shifted from one foot to the other, under all those tons of water and cement. Would the collapsing building bury him? he wondered. Or would he drown as the pipes and tanks all around him ruptured in a glassy inrush of seawater? Or would it be the sharks that got him, there in the aquarium's flooded underworld, thrashing now with all that life set loose?

Deep beneath my father, the Pacific Plate scraped against the ragged washboard of the North American Plate. The soil liquified. Concrete foundations fractured. Across the bay, the upper level of the Cypress Street Viaduct collapsed, killing forty-two drivers instantly; the survivors were cut from the rubble, quite literally: limbs sawed off and left behind under the shattered cement so that their bodies could be carried free. Above my dad, the sharks, with mouths like dinosaurs, agitated close against the edges of their tank.

And when my dad lived, when he made it out from below the miraculously unshattered tanks, he found he was the only one left in the aquarium. The atrium fluttered with decorations for the party that would never be. The building had never been quieter. He loaded the car with all the balloons and kites abandoned by the caterers, and he joined the slow parade of traffic spilling north across the Golden Gate Bridge. He arrived home with gifts for us all, the balloons pressing like hopeful faces against the windows of his battered little Honda.

For years, we took the aquarium kites flying on days when the wind was right. We spilled out of the car onto the beach and unspooled the long twine tethers and watched them go flying. We scampered, chasing the kites, through the dazzling fog and sun, which flared like curtains all down the coast from Bodega Bay clean through to San Francisco's skyline.

I often wonder what I would have done, alone and surrounded by that immensity of water, by that unthinkable load of concrete and glass and snappy-jawed fish. What would I have reached for when the rattling began? What is there to reach for, really, other than the upward-tugging red balloons, the forsaken kites, the unreasonable hope that when the shaking stops there will still be sky, and a bridge toward home, and the running feet of family on an open shore?

Salmon

1.

Thanksgiving, North Pacific coast. Just me, some salmon, and a man that I liked but maybe did not love. We poured mead, chopped garlic for citrusy kale salad. A salmon steak—caught fresh and trimmed neatly by some fishermen friends down the way—sizzled on the stove.

Its candied outer flesh would crisp against our teeth; the meat would dissolve, buttery, on our tongues. He would pump up some Jolie Holland, pull fresh bread out of the oven, and remind me that I didn't need to hold on if the distance was too hard. I'd never really know if that was a generosity or a request.

The next morning, we stepped off his boat and our shared cloud of breath intermingled for a moment in the cold air around our heads before dissolving upward and away from us. The dock was crisp with hoarfrost, and a goose had walked ahead, leaving tidy pigeon-toed prints that, in the residual warmth of the already distant bird, had begun, like the salmon, to melt.

2.

I go walking in my rubber boots over the crackled sidewalks, through pools of rainwater. Rectangular refrigerated containers belly up against cement cold storage buildings, and gulls rummage all down the docks, shaking beads of moisture from their beaks, pesking for fish scraps. The air smells like falling tide, and like spruce, and like dead salmon. At the side of the road: weather-gray telephone poles like totem poles. And in front yards: totem poles like telephone poles, moss-capped, straight-spined—a reminder that this world is older than the canneries, older than the ships that dock at them and unload their glittering catch.

South of here, in a cave I've often walked through—helmeted, flashlight panning—a ten-thousand-year-old human skeleton has just been pulled from a yellowy pile of bear bones. He was a young man, and a traveler—his stone tools came from far away. His bones, like books, tell the isotopic story of a man whose food was mostly fish. He had spent his lifetime pulling his food from the sea. So, too, had the bear who ate him—wandering up from the tideflats, glistening roe between its molars. A surprise encounter near the cave's entrance, where cool subterranean air sighed into the woods like breath between teeth. Quick knockdown, a second dinner. One salmon eater inside another. The bear crunched what it could but left a loose-toothed jaw, a snapped pelvis, three ribs to go cool in the cave over centuries.

Now, at the edge of town, mist lifts between the trees, and the sound of rain on moss dulls the dockside noise. A raven bristles and caws on a nearby branch, and it's hard not to think about old stories. Raven created this place and it's not really open for debate. You'd believe it, too, if you could hear the bell-tone voices of the bird-god's children. If you could see where their wing tips send fog curling like eddies in a creek. If you could watch their enormous black bodies—tall as toddlers—come swaggering up the road, their claws like soccer cleats, their eyes like bullets. They tilt their heads and watch me go by. Like the caveman's long-gone fingers, like the slicked fur of the grizzly that ate him, their feathers gleam with salmon oil.

3.

As I hitchhiked the long road between Anchorage and McCarthy, a carpenter on his way back from the hospital picked me up, showed me his bandaged hand where a bandsaw nearly severed his thumb clean off, and bought me some blueberry pie with his workers' comp cash at the only roadside truck stop within a hundred miles. Where my road diverged from his, he dropped me off, reached into the back of his car, and handed me a package of luminous salmon, freshly smoked in his own backyard. It should have been me—bedraggled, bumming rides at the side of the highway—who treated my benevolent chauffer to a gift, but still I accepted it graciously. I held it against my chest like treasure as he pulled away down the mountain road toward Valdez.

Later, too, when Alex moved into our house in Tucson, he offered a handful of smoked salmon, which had traveled with him all the way from Juneau after a summer working near Alaska's rain-grayed Inside Passage. And over old films one night, Jonathan—carver of ulu handles, videographer of Inuit dogsledders, Alaskan transplant here in the desert—got to talking with me about snow, about fish. He left the room and returned with a jar of salmon clutched in his hand, mailed all those miles from an old friend in the North. He offered me one fat piece after another. They fell apart in sugary splinters in my mouth: saltwater candy shared from some faraway sea.

And isn't the sharing the thing of it? When I visited an old shipmate at his inland home in eastern Oregon, we went fishing for salmon at Costco. We brined it—his uncle's recipe—in his kitchen, smoked it over wood chips in a vintage Little Chief smoker. We recalled Alaska together and laughed in the smoky backyard and cleared a patch of land for his garden while we waited for the fish to cure. When it was done, the candied flesh dissolved in our mouths. It tasted like sea, like molasses. If hospitality had a flavor, the salmon, I think, would have tasted like that.

4.

All summer, the salmon waited for rain. I got sunburned, even under the blanket of wildfire haze that settled low over Alaska, that stained the sky a bruisy orange, that would not lift. A rain forest without rain is an eerie place. Moss crackles underfoot. Mud dries into silty dust. Cedar forests go thirsty and send their oils aloft like war beacons so that, miles from shore and standing on deck as your ship cuts through deep water, you can smell their incense, sun-warmed, uncanny. The rain did not come. The salmon waited—waited for the creeks to fill, waited for enough water to scoot their way up under the spruce boughs, back up toward home.

Eleven million years ago, nine-foot-long salmon waited at the mouths of rivers along this same coast. Their four-hundred-pound bodies, like the bodies of modern-day salmon, transformed as they traveled upstream from the sea. They stopped eating. They metabolized their organs. Their skin lifted away in chunks and strips as freshwater flooded their salt-loving bodies, as their cells popped open like a thousand tiny bombs. Their jaws grew curved. Their teeth grew sharp, for tearing at one another as they fought upstream toward their wedding beds, which were also their graves.

Some years from now, coastal salmon farming operations—which leach parasites and antibiotics into wild waters and wild fish populations—will have moved indoors, into facilities that recirculate filtered water like giant aquariums. Here, in manufactured currents that flow at just the right pressure for ideal meat texture, the salmon of tomorrow will swim. They'll jostle behind thick glass viewing panes. They'll push their speckled bodies into the flow. Maybe one will pause for a moment, will drift backward in the downflushing current, knocking against mesh and tank and whirling neighbors. Maybe it will seek some familiar smell and, finding none, will muscle forward again, searching. If a salmon has no natal stream, toward what might it strain? Does movement lose meaning, or gain it, when there's no particular destination ahead?

My second summer in Alaska, a friend and I pulled our kayaks ashore on the final night of a paddling trip that had carried us along the edge of the ice age in Glacier Bay National Park. We had camped above a black sand beach and watched surf roll in as sheet after sheet of ice collapsed into the fjord from the face of a nearby glacier. At night, the katabatic wind running down off the ice rocked our tent with a chill persistence as steady as running water.

And now, we had begun our journey back down-bay. We'd pulled up our kayak and gone rummaging for firewood along the tide line. We walked across newly exposed rocks: a beach that had been covered by a glacier a 150 years ago when John Muir arrived by canoe to build his own campfire on the island across the way.

And next to us, in the raw gravel that was under forty stories of ice not so long ago, things rustled and flapped in a shallow stream. We stepped closer in the dusk to peer at the thin ribbon of water that cut across the beach near camp. Inside it, bodies thronged. Salmon, pushing forward. They weren't returning to their birth stream—they were taking new territory. Where ice had retreated, fish had come pioneering up-bay. Sometimes life loops back to where it began. Sometimes it hazards some new frontier. In the settling dark, at the edge of our campfire's circle of light, we found wolf tracks and bear tracks dimpling the rocky beach: life following life.

At summer's end that year, the rains came, knocking the smoke out of the air. The trees—the ones that had survived the drought—exhaled little clouds that rose like dragon's breath. The salmon, against odds, ran.

5.

Think of salmon milling upstream like lanes of traffic at rush hour. See them merge and glint.

Think of salmon, digested into shit and soil and spruce like recyclables in a furnace. All those shards re-rendered into glimmering whole new things: green and amber and clear as windows.

Think of their eggs, champagne pink and gummy like grains of split citrus. Hear them pop between bears' teeth: a thousand ripe desserts.

Think of salmon with their smell stripped by upstream copper mines, the ore in their blood like heavy fog—the kind that keeps ships from ever finding home.

Think of salmon with scales in colors of rock and gut and blush, their bodies tapestries shuttled through with movement, movement.

6.

And in the Northwest, where orcas eat salmon instead of seals, they get choosy. They like the fatty ones: the Chinook, the king salmon. Meriweather Lewis liked them, too—the most flavorful fish he'd ever tasted, he wrote in his journal as the Corps of Discovery ventured west at the turn of the nineteenth century.

When Lewis and Clark first tasted Chinook, given as a gift by Chief Cameahwait after the men had traversed the Bitterroots, they knew upon tasting it—that salty-sweet flesh, those flaky shingles of Pacific Ocean fat—that they had truly crossed the Continental Divide.

The resident pods of orcas all up and down the Inside Passage seek Chinook, and they teach their young to do the same. It doesn't serve them well, these days, as Chinook populations fall and fall—down 40 percent in most places—due to dams, and poor hatchery practices, and diseases seeping into the sea from coastal fish farms, and overfishing, and shoreline development that flattens gemlike stands of eelgrass and fills the brackish shallows where baby salmon need to shelter before they head to sea. The orcas—who have culture as we do, passed down from one generation to the next—grow skinnier year by year, as they swim in search of their favorite food. They ignore the lesser other salmon—silver, chum—streaking past them in the sea's green dapple, just as their grandmothers taught them.

But at the center of it all, the wonder—if you can afford a moment for wonder—is that orcas use their voices to identify the fish they love. They cast sound ahead of themselves and it pings back like sonar into a fatty organ called the melon, nestled just above the whales' jaws. The melon concentrates the echoes and sends signals to the orcas' brains, clear as photographs, that help guide them toward prey. So sophisticated is their echolocation that they can tell individual salmon species apart, just by the subtly different way their bodies reverberate.

And think what an astonishing thing it is, despite the things that are unhinging all around the orcas, and around us, in the periphery—too lightless and too huge to really apprehend, even with eyes as sharp as ours. Consider the miracle of it: To see something with sound. To know a thing with your body.

Fortune Fish

IN THE GLEN ELLEN Village Market, perched at the corner of the one main road that ran through my little hometown, Raegine Africa sold long candy sticks in jars, striped like barber poles, burgundies and blues and toffee browns. She sold craggy bits of rock candy, too— otherworldly neon cave crystals that melted pink and slow on the tongue. She sold antiques, and china dolls, and little clothespins with washers glued to them in a bin labeled HILLBILLY WASHER AND DRYER, and stuffed rabbits with sweet faces stitched in thread. And, in the back of the store, on a shelf low enough for small hands to find, she kept a basket of tissue-thin envelopes, stamped red and white, each containing a translucent red fish.

A Fortune Teller Miracle Fish came out of its envelope as flat and fine as photographic film. Its body showed, faintly, all the markings that real fish have: gills, eyes like coins, scales inked into its red body with slightly darker shades of red so that, held up to the light, it was an etched lantern, pinkish and luminous and alive with portent. Placed across the palm, the fish would move. It would bend and thrash like a real fish pulled free of the water. Kneeling in the back of the mercantile, I would wait breathlessly for the fish to settle into position, so that I could decipher its message using the key printed on the back of the envelope. If the fish's sides curled inward, it meant that I was fickle, but if the body closed entirely into a tight loop, it meant, instead, that I was passionate. Where the fish bent, how it flipped in my palm, whether its head or its tail flapped upward, all were omens for the future—omens for indifference, for affection, for independence.

If magic could happen anywhere, it was in the back of Raegine Africa's little shop—fragrant with sugar, flashing with antiques that could have belonged to wandering wizards or to hoodoo rootworkers—and so I trusted that the little fish knew things that I did not.

Looking back, I suppose the fish were no more accurate than fortune cookies or arcade-prize mood rings. Their bodies were made out of sheets of sodium polyacrylate, after all: the same absorbent salt that manufacturers use to make disposable diapers. Sodium polyacrylate, also called waterlock, latches onto moisture. The cellophane-thin fortune fish plucks wetness from a person's palm and, depending on how sticky different parts of our hands are—or which part of the fish touches down first—the fish will contort and wiggle and arrive at some new vision for the bearer's future. That's the science of it.

But—as collapsed glacial ice, dense with centuries of compressed air, sizzles like Rice Krispies in the sea-size bowl where I paddle; as wildfires in my hometown send a married couple tumbling into a nearby pool, where they hold each other all night in the black smoke; as pandemic comes knocking on each of our doors; as I loosen the buttons on a lover's shirt, hoping this time to find something tender beneath; as vaquitas, the last of their kind, go calling for one another in the gray-blue gulf, go not finding one another, go gone—wouldn't it be nice to have a fortune fish, now? A crinkly red omen, a back-shelf mercantile oracle, an assurance (graspable and close, there in the scoop of your hand) that catastrophe won't be the place where this all ends?

Lay it across your hand, let the warm wetness of your body send it curling. Watch the truth unfurl in your palm. See how its ragged little body bends and bends and bends, spelling out love.

Convergences

ONE SUMMER, AWAY FROM the orcas that are growing thin from too few fatty salmon, away from coastlines gooey with the melting bodies of sea stars—forty different species, all disintegrating like cotton candy gone wet and no one knows why—I go walking in the mountains with Betsy. We plan to hike from a glacier's edge across a mountain range and down into a neighboring glacier valley in the Wrangell Mountains, a wilderness only paralleled in size by Antarctica. We plan to leave our world for a while and enter a different one altogether. At the highest point, our route will carry us through Oz: a terraced emerald wonderland of alpine lakes and heather. We bring a few days' provisions, a big black-and-white dog named Buddha, and a satellite radio in case our pilot needs help finding us on the other side.

The Wrangell Mountains are made out of old rock. Betsy and I start our hike at the Fosse, a flat shelf of stones pushed high and left behind by the melting Kennicott Glacier below us. The rocks came from the surrounding peaks: greenstone, limestone, blue copper ore. Rocks split by ice, plucked by glaciers, borne like jars on a slow-sliding factory conveyor belt.

We clatter over rocks that formed when the world was young—rocks that, three hundred million years ago, were part of a renegade island arc that drifted all on its own at sea while the rest of the world's land was bunched together in Pangaea, the arid supercontinent. The lonely archipelago was called Wrangellia. All those millions of years ago, layers of sea life sifted like flour along Wrangellia's flanks: sharks, squids, coiled ammonites with shells like wild sheep's horns. They stacked and crystallized into limestone, into fossil—a salty marine layer cake, of sorts. We move across the crumbs.

We slide down a steep hill woven tight with blueberry bushes and scrubby alders. Our backpacks catch on low branches. Buddha stops to nibble on berries, his white whiskers alive among the smells, the punchy flavor. Ahead is a broad valley where, just a few days ago, a massive lake lapped, rimmed by rock on three sides and by glacier on its far western edge. Over the weekend, warm weather had melted out a glacial plug and, like a bathtub, the lake swirled dry. Now, the empty valley glistens with stranded icebergs. They drip. They shimmer in the sun. We weave through them as if walking through a city: muddy pavement sticking to our soles, skyscrapers leaning, dog scrambling ahead.

As tectonic plates got restless, Wrangellia got hot. Huge eruptions sent basalt pillowing across the land, sizzling into the sea. We've seen similarly big eruptions within human memory: in June of 1783, 130 craters opened along a fifteen-mile fissure in southern Iceland. Lava fountains spat four thousand feet high and demolished twenty villages. Heat from the eruptions carried gas ten miles into the sky. A blood-colored sun rose on Norway, and a poisonous cloud drifted across the globe, bringing thick fog that stayed boats at anchor. That year, hundreds of thousands of people died in the noxious haze. Thunderstorms spat hailstones big enough to fell whole herds of livestock. Floods rampaged across Germany. That winter, more snow fell in New Jersey than ever recorded before or since. Ice floes clicked and drifted across the Gulf of Mexico. In France, the ensuing years of bad weather planted seeds shaped like drought, seeds shaped like poverty, seeds that would grow into the French Revolution.

A single volcanic event can send global climate reeling. Among the islands that would someday become the Wrangell Mountains, eruptions as huge as the one in Iceland happened every other year, for hundreds of thousands of years. The basalt cooled in miles-deep layers.

And following the eruptions came rain as the world had not seen in a very long time. A million years of rain. Rain that teased humid jungles up out of Pangaea's cracked earth, rain that sent new rivers following new paths across the newly dividing continents, rain that life (laid flat in the earthshaking Permian-Triassic extinction twenty million years before) took advantage of, springing up like an evolutionary jack-in-the-box. Dinosaurs, still young on the earth, exploded in population and diversity. The rain gave birth to the ancestors of *Triceratops* and *Tyrannosaurus rex*. They squelched and thundered across the wet earth, they zigzagged through buzzing rain forests, they shook bright beads of moisture from their feathery skins.

And during this time, during the Great Wet, the oldest known mammal fossils appeared, stamped in mud. A gift from Wrangellia: a million years of rain, and the origin of our species.

Beyond the grounded icebergs, at the valley's far end, our climb toward Oz begins. We carve an upward diagonal path for ourselves across talus. Fossils come loose underfoot. I have to bend to spot them, but Betsy, who has lived a long time in these mountains, has an eye for rocks that are different from other rocks. She hands me fragments of ammonites, embedded in stone like dark snails. I turn a dusty rock in my palm and trace the little scalloped edges of a shell. Though the animals lived a long time ago, the shapes feel familiar against my fingers: They could easily be the coiled folds of a nautilus, the ribbed curve of a clam.

There are no trails to follow in the Wrangell Mountains. We find our way as best we can, pushing higher among gappy crevices and slopes ragged with scree. There's a route-finding saying in the Wrangells, Betsy tells me: *When in doubt, go high*. But the higher we go, the sheerer the mountainside gets. The flinty fossils slide like playing cards, and for every step we take up and forward, we go skating downhill, too. Betsy is athletic, confident, crossing slopes in long, taut-calved strides. I falter and slip and step gingerly from one unconvincing purchase to the next. We cross ravines. We cling to stringy alder branches, hoping they will hold our weight until we find our next footing. And eventually, constricted by deep cuts in the hillside, we give up on the high route. We pick our way downhill toward the creek valley below. Like time travelers, we descend through the stony epochs, rocksliding among loose strata, looking for a toehold somewhere between the Permian and the Triassic.

We set up camp. Buddha guards us, belly against stone, ears swiveling for bears.

In the Panthalassa Ocean, where Wrangellia formed, water stretched in every direction. Inside the scoop of Pangaea was the Tethys Ocean—smaller, shallower, warmer than the outside water. Most of what we know about Pangaea-era sea life comes from the Tethys and the tropical coasts that surrounded that supercontinent, exposed now in fossil shelves in Texas, in Egypt, in Pakistan. But the Panthalassa remains elusive in our collective imagination. All that oceanic stuff—grit and shells and bone from that glittering open sea—got tucked deep beneath the earth's mantle, or pushed up in isolated places like the densely glaciated Wrangell Mountains, where paleontologists seldom ventured. And so the enormous Panthalassa Ocean and the life it harbored have remained mysteries to science. Where Betsy and I sleep, on the flanks of Hidden Creek Valley, our bodies might overlay the bodies of entirely unknown things, awaiting exhumation. Muscular fish, built for wide water. Fat prehistoric sharks, their ribs like open birdcages below the earth. Whole corpses petrified in seafloor sand, spread like tarps beneath our tents, beneath Alaska's dusky and never-dark night sky.

In the morning, we leave our bags behind at camp and explore the far side of our valley. We've seen a hole next to a waterfall across the way and we think we might like to climb inside it, if we can reach it. The cave sits dark and high on the mountainside, near the seam where basalt from the long-ago Wrangellia eruptions (metamorphosed into marble-like greenstone now, by the warp of time and pressure) meets the younger limestone above it. We cross the creek on the valley floor: shock of meltwater against bare ankles. We climb along the edge of the waterfall, gripping roots and mossy tufts. The billowing vapor is cool on our necks. Buddha follows where we climb, leaping across the waterfall with us, balancing on mist-slicked rocks, but when we get to the entrance of the cave, he sits tight. He does not whine, but he does not budge. *Smart dog*, I think.

Betsy gives him a nuzzle, and we click on our head lamps, and Buddha waits at the entrance while we slide upward into the dark.

We move toward the back of the cave and, where the walls close in, we discover that the cave does not end but instead narrows, leading upward. Doughy folds of rock frame the entrance to this tighter opening, like curtains around a door. We climb up through the passageway.

Inside, it is a miracle, a dazzlement. The ceiling of the cave drips with pale crystals, luminous as moons. The edges crackle with black, faceted gems. Our head lamps send light dimly through the translucent minerals. Every surface glimmers, sparks. These are spar crystals—pulled up by moving water, latticed out of old island-chain minerals: calcite, gypsum, quartz. It is like moving through a supersize geode. We could be in another world, another time altogether. I imagine whole Panthalassa skeletons fossilized somewhere within the glittering walls, fins spread, snaggly mouths with rocky teeth, caught mid-chomp as in some Triassic Pompeii.

As we follow the chamber deeper into the mountainside, the sound of moving water rises around us. We round a bend and find a stream moving cooly across the diamond-like floor—blue light pooling where our beams move. Do old things live here, underwater, under rock? Vestiges, hangers-on from long ago? I imagine cave fish—the pale ones, the ones with blush-pink patches on their cheeks where their insides shine through their gills. The blind ones, the ones whose eyes melted gone and whose fins grew long to conserve energy. Do they lurk here? How would they have arrived? It would be hard to know: Cave fish don't all descend from a pinpointable ancestor. When fish find themselves in caves, they adapt. One generation at a time, their bodies change. Fish from dozens of families have arrived at their own version of cave-fishiness, each along their own distinct evolutionary routes. Catfish and gobies, carps and loaches, snakeheads and glass knifefish—all these gave birth, independently, to cave fish around the world.

Through these different lineages of cave fish (scaleless, sightless, tender bodied—unable to cross salty seas on their own), scientists can track the split and drift of whole continents—a particular cave fish species that moved through buried water in Laurasia, for example, might be findable now in Siberia, Nepal, Nova Scotia: the crumbs of land that pulled away from that long-ago landmass. The world's cave fish show what was once connected.

I peer into the moving water. No life to be seen: just chime and plink, as water seeks a path through this old mountain. All around us, the gems in the walls and the ammonite fossils stacked in the strata and the maybe-fish in these underground channels are what my mother simply calls *repeating patterns*, which she loves: the familiar sacred geometry of a nautilus, the comforting repeating latticework in crystals, the emergence of cave fish after cave fish after cave fish—in new caves, in new fish families, on new continents.

In biological circles, this is called convergent evolution: when life in separate but alike places takes on similar characteristics to suit similar environments. Spiny North American cacti and plump, thorny African euphorbias. Echolocation among bats and among dolphins. The poky backs of echidnas and hedgehogs.

Repeating patterns. Known shapes. I am thinking about these things because so little feels familiar, or sure, or perseverant in the world that I left behind when I hitchhiked twenty hours into the Wrangell Mountains to visit Betsy. In the Sea of Cortez, where I work in the winters, only about ten stocky, shiny-backed little porpoises called vaquitas are still alive. There may be none left in another year or two; maybe they'll join the ten thousand or so species—whole species—that go extinct every year now in the epoch that some scientists are beginning to call the Anthropocene. In coastal Alaska, where I work in the summers, I guide people across long miles of silt and gravel—raw earth left behind by quickly thinning glaciers. Ours are likely the first human feet to travel this newly scoured ground. Patterns distorted, disrupted.

But south of here, in isolated lakes scattered across British Columbia, three-spined sticklebacks—big eyes, bodies splotched with teal and red—are making new patterns. Or old ones, maybe. What began as one species of stickleback has split into two. Some of them have adapted to forage along the mucky lake bottoms. Others have adapted to swim out in the open water, in pursuit of little invertebrates to eat. In each lake, this process has happened independently, separately. That is, sometime in the geologically brief last twelve thousand years, in several different lakes in British Columbia, a single species of fish evolved into two different species, in isolation from the identical evolutionary gears turning in neighboring bodies of water. Many different lakes. In each: first one species, then two. They are some of the youngest species on earth.

Recently, scientists have begun pulling sticklebacks out of their homes and dropping them into artificial ponds intended to simulate the wild lakes that they come from. In each pond, the same thing happens: Two species grow out of one. Some sticklebacks take to the soft lake beds. Others stick to the higher currents. Their bodies change. They stop interbreeding with one another. This happens not over millions or thousands or even hundreds of years but over the course of just three or four years. Evolution faster than Charles Darwin ever could have predicted. You might call it divergence—one species splitting into two. But from a big-picture perspective, it is also a convergence: In separate ponds, analogous species evolve to suit the conditions they're given. A bottom-dwelling stickleback from one pond might very closely resemble its bottom-dwelling neighbor in the next pond over, though they evolved separately. Their alikeness is dictated by the alikeness of their environments. Convergent evolution in action: separate but similar places steering life toward similar characteristics. Time and again, where there is a lake bed, fish will take on just the right shape to make use of it. Patterns repeated.

The Canadian sticklebacks are a speculative evolutionist's dream. If, time after time, similar environments breed similar shapes of life (coiled shells in deep currents, pale fish in dark caves, pairs of lake-bound species to suit pairs of adjacent habitats), why wouldn't that extend beyond the epoch we're living in, why wouldn't that extend beyond—even—our little planet? Convergent evolution inspires us to imagine worlds outside our own. Though none of the other planets in our solar system have the right conditions for life as we know it to evolve, they are our nearest neighbors, and so we look to them, and extrapolate.

Europa, Jupiter's smallest moon, churns with a salty, hidden ocean. The tides and storms on that vast body of water split the ice on Europa's surface, knocking the frozen slabs around like puzzle pieces on a table. We cannot see the ocean, but we can trace its presence in the fractures in the ice. If another, warmer, planet had such an ocean, what other circumstances would need to align for cells, for sponges, for many-finned marine creatures that look familiarly fishy?

On Mars, enormous buried glaciers ooze out from the edges of mountains and cliffs. They are insulated under deep blankets of shattered rock. Analysts for the NASA Mars Reconnaissance Orbiter wonder if they might be vestiges of an ice sheet that would have blanketed the planet during some long-ago ice age. Maybe someday these buried rivers of ice will melt away entirely, as ours here on Earth are beginning to do. Maybe the broad valleys that the glaciers scooped out will fill with meltwater, will become lake beds, birthplaces for a species—or two—of proto-stickleback.

Of course, the atmosphere on Mars is too thin for life, but under its surface lie lava caves. We cannot see below the planet's crust with satellites, but we can spot the entrances, the skylights—wide holes, collapsed shafts. Underground, perhaps, are mazes of empty veins, echoing chambers big enough for spaceships. Though Mars has no liquid water anymore, billions of years ago it likely sloshed with oceans and with underground aquifers. Imagine it: the dark water moving beneath stone, teasing crystals up out of volcanic minerals, carving the caves into deep and intricate tunnels. Imagine the warmth rising up from the planet's core, the barrier that the cave ceilings provide against the thin and unforgiving sky, the holy and sudden stir of microbial life in those lightless places. Imagine, many hundreds of millions of years later, on Mars or elsewhere in the cosmos, the fishlike things that might arrive in those watery tunnels. The way their skin, away from the sun's scorch, grows pale. The way that their eyes, blind in the blackness, melt away. The way that they fan their lengthening fins and brush their smooth bodies against their gemlike aquarium walls.

Convergent evolution is like a prayer bead, a promise. Beyond earthly progressions of continent and drift, disaster and imbalance—even beyond Earth itself, perhaps—familiar patterns return, persistent as a hungry dog.

On the far side of Oz, we scramble, days later, into the valley where our pilot will meet us. My heels are raw and fat with blisters. Fine glacial dust rises among thin spruce and the air is warm. We push through the edge of the woods and onto the gravelly riverbank, wandering north in search of our meeting spot. We take off our shoes and let the sand sink between our toes, let the stones roll like knuckles across the beds of our aching feet.

We are not the only ones moving across the valley today: Wolf tracks lead north, too. We follow them. When the day gets hot and the sky is still quiet, we set down our shoes at what we hope is the right landing spot. Buddha curls up to snooze in the sun. We peel away our sticky shirts, our dusty pants. We walk out toward the middle of the valley, where eddies of gray glacier water have gathered in still pools, away from the downward pull of the current. They are still icy, but warmer than the flowing water beyond them. Silt and mountain minerals have settled out, and the water in the pools is turquoise and translucent. Naked, we lower our bodies into the shallows, yelping from the cold. We lay back, we laugh, we spread our arms open like stars. Behind closed eyes, my lids flare with afterimages of the landscapes we have moved through: dripping bergs, the embrace of the cave, the dizzy green sprawl of Oz.

Somewhere, in some other solar system, a womanlike creature has stripped her clothes away, too. She pads out across the banks of some Earthlike river, lowers her body into the current. Below her, in caverns jagged with crystals, fishlike things stir in subterranean pools. Above her, in the walls of the mountains, the old bodies of shelled creatures from a long-ago sea lie buried, each one a tight and perfect coil, geometric, sacred. On her planet, and on others all across galaxies, a thousand lakes give life to a thousand nearly identical species of scaly swimmers—teal and red bellies, sharp spines cutting up through their diaphanous fins, bodies shaped perfectly to suit the lake-bed muck. Is it self-indulgent to believe that things similar to us might come to exist at other times, in other places, just as we have come to exist out of microbial murk, out of finely boned ancestors, out of catastrophe? It is self-indulgent not to? After disaster: a million years of rain. After disruption: convergence.

Here on the banks of the Lakina River, the approaching hum of a Cessna bush plane rattles the rims of the ancient mountains. Behind us on the beach, among tufty ice age flowers, our tracks merge with the wolves' tracks. The prints fit themselves into the soft shape of the riverbed. They move forward and circle back, returning, converging into familiar patterns.

Acknowledgments

Thank you, thank you:

To Darcy Frey, for first showing me what nonfiction writing can be, all those years ago when I first wrote about goldfish.

To my initial readers of this collection: Chris Cokinos, Aurelie Sheehan, Jennifer Sinor, Alison Hawthorne Deming, Ander Monson, and to my cohort at the MFA program at the University of Arizona, who tolerated my insistence that landscapes and creatures can be the main characters in a story.

To Amary Wiggin and Leslie Beach, for your unwavering support as friends and fellow writers in a world where words aren't always valued.

To Mom, for reminding me to omit needless words—and for teaching me never to turn my back to the sea.

To Jane, for your fearless fieldwork: dating men with "fish pics" in their profiles, all in the name of research.

To the professors who have taught me about fish and their watery worlds: Karel Liem and Robert Woollacott at Harvard University, Jim Carlton at the Williams-Mystic Maritime Studies program, and Peter Reinthal at the University of Arizona.

To my brother, who has always looked out for me. Being in this world is hard and beautiful, and you put in the work. Your recovery inspires me beyond words.

To the founders and editors at *River Teeth*, for publishing my very first shaky piece of creative nonfiction a decade ago, and now for celebrating my work with the River Teeth Literary Nonfiction Book Prize: the most astonishing full circle. And to Beth Nguyen, for your perceptive and generous attention to my manuscript.

To the fish, for teaching me about survival, transformation, buoyancy, and joy.

To the places that have held me as I've lived and relived the stories in this collection: Southeast Alaska; Tucson, Arizona; the sea surrounding Hawai'i; the islands of the Sea of Cortez; the Sitka Center for Art and Ecology residency; the Spring Creek Ranch Writer's Residency; the Breadloaf Environmental Writers' Conference; the Montello Foundation residency; the Wrangell Mountains Writing Workshop; and Louis, who looked past the red flags in these essays and who has become for me a homeplace. I love you more than all the fish in the sea.

And, of course, to Dad: for your love that knew no limits, for your infectious delight in marine life, chanteys, and the powerful Pacific, and for staying so close and so alive in my heart even now, so long after the sea swallowed you whole.